ON THE PRINCIPLES
OF ELEMENTARY QUANTUM MECHANICS

N.V. VAN DE GARDE & CO'S DRUKKERIJ, ZALTBOMMEL

ON THE PRINCIPLES
OF ELEMENTARY QUANTUM MECHANICS

PROEFSCHRIFT

TER VERKRIJGING VAN DEN GRAAD VAN
DOCTOR IN DE WIS- EN NATUURKUNDE
AAN DE RIJKS-UNIVERSITEIT TE UTRECHT
OP GEZAG VAN DEN RECTOR-MAGNIFICUS
Dr. W. P. J. POMPE, HOOGLERAAR IN DE
FACULTEIT DER RECHTSGELEERDHEID,
VOLGENS BESLUIT VAN DEN SENAAT DER
UNIVERSITEIT TEGEN DE BEDENKINGEN
VAN DE FACULTEIT DER WIS- EN NATUUR-
KUNDE TE VERDEDIGEN OP DINSDAG 29
OCTOBER 1946, DES NAMIDDAGS TE 3 UUR

DOOR

HILBRAND JOHANNES GROENEWOLD
GEBOREN TE MUNTENDAM

SPRINGER-SCIENCE+BUSINESS MEDIA, B.V.
1946

ISBN 978-94-017-5719-5   ISBN 978-94-017-6065-2 (eBook)
DOI 10.1007/978-94-017-6065-2

Promotor: PROF. DR. L. J. H. C. ROSENFELD

CONTENTS

5. Quasi-distributions.

ON THE PRINCIPLES
OF ELEMENTARY QUANTUM MECHANICS

Summary

Our problems are about

α the correspondence $a \longleftrightarrow \mathbf{a}$ between physical quantities a and quantum operators $\mathbf{a}$ (quantization) and

β the possibility of understanding the statistical character of quantum mechanics by averaging over uniquely determined processes as in classical statistical mechanics (interpretation).

α and β are closely connected. Their meaning depends on the notion of observability.

We have tried to put these problems in a form which is fit for discussion. We could not bring them to an issue. (We are inclined to restrict the meaning of α to the trivial correspondence $\mathbf{a} \to a$ (for $lim\ \hbar \to 0$) and to deny the possibility suggested in β).

Meanwhile special attention has been paid to the measuring process (coupling, entanglement; ignoration, infringement; selection, measurement).

For the sake of simplicity the discussion has been confined to elementary non-relativistic quantum mechanics of scalar (spinless) systems with one linear degree of freedom without exchange. Exact mathematical rigour has not been aimed at.

1. Statistics and correspondence.

1.01 *Meaning*. When poring over

α the correspondence $a \longleftrightarrow \mathbf{a}$ between observables a and the operators $\mathbf{a}$, by which they are represented in elementary quantum mechanics,

β the statistical character of elementary quantum mechanics

(we need α for β), we run a continuous risk of lapsing into meaningless problems. One should keep in mind the meaning of the conceptions and statements used. We only consider

1

M_o: *observational meaning*, determined by the relation with what is (in a certain connection) understood as observation,

M_f: *formal meaning*, determined with respect to the mathematical formalism without regard to observation.

Only M_o is of physical interest, M_f is only of academic interest. Dealing with M_f may sometimes suggest ideas, fruitful in the sense of M_o, but may often lead one astray.

1.02 *Quantization*. Very simple systems suffice for demonstrating the essential features of α and β. In elementary classical point mechanics a system is described by the coordinates q of the particles and the conjugate momenta p. We only write down a single set p,q, corresponding to one degree of freedom. Any other measurable quantity (observable) a of the system is a function $a(p,q)$ of p and q (and possibly of the time t). The equations of motion can be expressed in terms of P o i s s o n brackets

$$(a,b) = \frac{\partial a}{\partial p} \frac{\partial b}{\partial q} - \frac{\partial a}{\partial q} \frac{\partial b}{\partial p}. \tag{1.01}$$

When the same system is treated in elementary quantum mechanics, the (real) quantities a are replaced by (H e r m i t i a n) operators $\mathbf{a}$, which now represent the observables. In the equations of motion the P o i s s o n brackets (1.01) are replaced by the operator brackets

$$[\mathbf{a},\mathbf{b}] = \frac{i}{\hbar}(\mathbf{ab} - \mathbf{ba}) \; (\hbar = \frac{h}{2\pi}, \; h \; \text{P l a n c k's constant of action}). \tag{1.02}$$

Problem α_1 is to find the correspondence $a \to \mathbf{a}$ (other problems α are stated further on).

1.03 *Statistical character*. The statements of quantum mechanics on observations are in general of statistical character. Problem β is whether the statistical quantum processes could be described by a statistical average over uniquely determined processes (statistical description of the 1st kind, type S^1) or not (statistical description of the 2nd kind, type S^2). The observability of the uniquely determined processes may be required (proper statistical description, type S_o) or not (formal statistical description, type S_f). (Classical statistical mechanics, e.g. are properly of the 1st kind, type S_o^1).

1.04 *Transition operator*. Before going on we have to deal for a moment with the operators and the wave functions.

The H e r m i t i a n operators **a** form a non-commutative ring. The normalized elements (wave functions) of (generalized) H i l b e r t space on which they act from the left are denoted by φ_μ, the adjoint elements on which they act from the right are denoted by $\varphi_\mu^\dagger$. Unless otherwise stated the inner product of $\varphi_\mu^\dagger$ and φ_ν is simply written $\varphi_\mu^\dagger \varphi_\nu$. The outer product of $\varphi_\mu^\dagger$ and φ_ν defines the transition operator

$$\mathbf{k}_{\nu\mu} = \varphi_\nu \varphi_\mu^\dagger, \quad \mathbf{k}_{\nu\mu}^\dagger = \mathbf{k}_{\mu\nu}. \tag{1.03}$$

Take a complete system of orthonormal wave functions φ_ν. The orthonormality is expressed by

$$\varphi_\mu^\dagger \varphi_\nu = \delta_{\mu\nu}, \tag{1.04}$$

the completeness by

$$\sum_\mu \varphi_\mu \varphi_\mu^\dagger = \mathbf{1}. \tag{1.05}$$

In continuous regions of the parameter μ the W e i e r s t r a s z δ-symbol must be replaced by the D i r a c δ-function and the sum by an integral. (1.04) and (1.05) show that every (normalizable) function φ can be expanded into

$$\varphi = \sum_\mu f_\mu \varphi_\mu \text{ with } f_\mu = \varphi_\mu^\dagger \varphi. \tag{1.06}$$

$\mathbf{k}_{\nu\mu}$ and $\mathbf{k}_{\nu\mu}^\dagger$ transform $\varphi_{\mu'}$ and $\varphi_{\mu'}^\dagger$ according to

$$\mathbf{k}_{\nu\mu}\varphi_{\mu'} = \varphi_\nu \delta_{\mu\mu'} \text{ and } \varphi_{\mu'}^\dagger \mathbf{k}_{\nu\mu}^\dagger = \delta_{\mu'\mu}\varphi_\nu^\dagger \tag{1.07}$$

(that is why they are called transition operators). (1.04) gives

$$\mathbf{k}_{\mu\nu}\mathbf{k}_{\nu'\mu'} = \mathbf{k}_{\mu\mu'}\delta_{\nu\nu'}. \tag{1.08}$$

In particular $\mathbf{k}_{\mu\mu}$ and $\mathbf{k}_{\nu\nu}$ are for $\mu \neq \nu$ orthogonal projection operators (belonging to φ_μ and φ_ν respectively).

The trace of an operator **a** (resulting when **a** acts towards the right upon itself from the left, or opposite; when it bites its tail) is (according to (1.05)) defined by

$$Tr\mathbf{a} = \sum_\mu \varphi_\mu^\dagger \mathbf{a} \, \varphi_\mu. \tag{1.09}$$

(Because the right hand member is invariant under unitary transformations of the φ_μ, this definition is independent of the special choice of the complete orthonormal system of φ_μ). This gives

$$Tr(\mathbf{k}_{\nu\mu}\mathbf{a}) = \varphi_\mu^\dagger \mathbf{a} \, \varphi_\nu. \tag{1.10}$$

(1.04) and (1.05) can be written

$$Tr\mathbf{k}_{\nu\mu} = \delta_{\nu\mu}, \tag{1.11}$$

$$\sum_\mu \mathbf{k}_{\mu\mu} = \mathbf{1} \tag{1.12}$$

and further imply

$$Tr(\mathbf{k}_{\mu\nu}\mathbf{k}_{\nu'\mu'}) = \delta_{\mu'\mu}\delta_{\nu\nu'}, \tag{1.13}$$

$$\sum_{\mu,\nu} \mathbf{k}_{\nu\mu}Tr(\mathbf{k}_{\mu\nu}\mathbf{a}) = \mathbf{a} \text{ (for every } \mathbf{a}). \tag{1.14}$$

(1.13) and (1.14) show that every operator $\mathbf{a}$ (with adjoint $\mathbf{a}^\dagger$) can be expanded into

$$\mathbf{a} = \sum_{\mu,\nu} \alpha_{\nu\mu}\mathbf{k}_{\mu\nu} \text{ with } \alpha_{\nu\mu} = Tr(\mathbf{k}_{\nu\mu}\mathbf{a}). \tag{1.15}$$

$\alpha_{\nu\mu}$ is the matrix element (1.10) of $\mathbf{a}$ with respect to φ_ν and φ_μ. It follows further that if $Tr(\mathbf{ac}) = 0$ for every $\mathbf{a}$, then $\mathbf{c} = 0$ and therefore (1.14) is equivalent to

$$\sum_{\mu,\nu} Tr(\mathbf{k}_{\nu\mu}\mathbf{b}) \, Tr(\mathbf{k}_{\mu\nu}\mathbf{a}) = Tr(\mathbf{ab}) \text{ (for every } \mathbf{a} \text{ and } \mathbf{b}). \tag{1.16}$$

Further

$$Tr(\mathbf{ab}) = Tr(\mathbf{ba}). \tag{1.17}$$

When $\mathbf{a}$ is a H e r m i t i a n operator

$$\mathbf{a}^\dagger = \mathbf{a}, \quad \alpha^*_{\nu\mu} = \alpha_{\mu\nu} \tag{1.18}$$

(the asterik denotes the complex conjugate), the system of eigen-functions φ_μ with eigenvalues a_μ

$$\mathbf{a}\varphi_\mu = a_\mu\varphi_\mu \tag{1.19}$$

can serve as reference system. In this representation (1.15) takes the diagonal form

$$\mathbf{a} = \sum_\mu a_\mu \mathbf{k}_{\mu\mu}. \tag{1.20}$$

1.05 *Statistical operator* [1]). The quantum state of a system is said to be pure, if it is represented by a wave function φ_μ. The statistical operator of the state is defined by the projection operator $\mathbf{k}_{\mu\mu}$ of φ_μ. We will see that the part of the statistical operator is much similar to that of a statistical distribution function. The most general quantum state of the system is a statistical mixture of (not necessarily orthogonal) pure states with projection operators $\mathbf{k}_{\mu\mu}$ and non-negative weights k_μ, which are normalized by

$$\sum_\mu k_\mu = 1. \tag{1.21}$$

(In some cases the sum diverges and the right member actually should symbollically be written as a δ-function). The statistical operator of the mixture is (in the same way as it would be done for a distribution function) defined by

$$\mathbf{k} = \sum_{\mu} k_{\mu} \mathbf{k}_{\mu\mu} \qquad (1.22)$$

and because of (1.21) normalized by

$$Tr\mathbf{k} = 1. \qquad (1.23)$$

(we will always write 1 for the right member, though in some cases it actually should be written as a δ-function). For brevity we often speak of the state (or mixture) $\mathbf{k}$.

An arbitrary non-negative definite normalized H e r m i t i a n operator $\mathbf{k}$ ($Tr\mathbf{k} = 1$) has non-negative eigenvalues k_{μ}, for which $\sum_{\mu} k_{\mu} = 1$ and corresponding eigenstates with projection operators $\mathbf{k}_{\mu\mu}$. Therefore $\mathbf{k}$ can according to (1.20) be expanded in the form (1.22) and represents a mixture of its (orthogonal) eigenstates with weights given by the eigenvalues.

The statistical operator $\mathbf{k}_{\mu\mu}$ of a pure state is from the nature of the case idempotent ($\mathbf{k}_{\mu\mu}^2 = \mathbf{k}_{\mu\mu}$). If on the other hand an idempotent normalized H e r m i t i a n operator $\mathbf{k}$ is expanded with respect to its eigenstates $\mathbf{k}_{\mu\mu}$ with eigenvalues k_{μ}, we get

$$\mathbf{k}^2 = \mathbf{k}, \ k_{\mu}^2 = k_{\mu}; \ Tr\mathbf{k} = 1, \ \sum_{\mu} k_{\mu} = 1, \qquad (1.24)$$

so that one eigenvalue k_{ν} is 1, all other are 0. Then $\mathbf{k}$ is the projection operator of the pure state φ_{ν}

$$\mathbf{k} = \mathbf{k}_{\nu\nu}. \qquad (1.25)$$

Therefore pure states and only these have idempotent statistical operators.

Suppose the normalized statistical operator $\mathbf{k}$ of an arbitrary quantum state is expanded in some way into other normalized (but not necessarily orthogonal) statistical operators $\mathbf{k}_r$ with non-negative weights k_r

$$\mathbf{k} = \sum_{r} k_r \mathbf{k}_r; \ \ k_r \geqslant 0. \qquad (1.26)$$

This gives

$$\mathbf{k} - \mathbf{k}^2 = \sum_{r} k_r \ (\mathbf{k}_r - \mathbf{k}_r^2) + \tfrac{1}{2} \sum_{r,s} k_r k_s (\mathbf{k}_r - \mathbf{k}_s)^2. \qquad (1.27)$$

If we expand with respect to pure states $\mathbf{k}_r$ ($\mathbf{k}_r^2 = \mathbf{k}_r$), (1.27) becomes

$$\mathbf{k} - \mathbf{k}^2 = \tfrac{1}{2} \sum_{r,s} k_r k_s (\mathbf{k}_r - \mathbf{k}_s)^2. \qquad (1.28)$$

This shows that $\mathbf{k} - \mathbf{k}^2$ is a non-negative definite operator. If the given state is pure ($\mathbf{k}^2 = \mathbf{k}$) all terms at the right hand side of (1.27) (which are non-negative definite) must vanish separately. For the terms of the first sum this means that all states $\mathbf{k}_r$ with non-vanishing weight ($k_r > 0$) must be pure, for the terms of the second sum it means further that all these states must be identical with each other and therefore also with the given state ($\mathbf{k}_r = \mathbf{k}$). The given state is then said to be indivisible. If the given state is a mixture, $\mathbf{k} - \mathbf{k}^2$ must be positive definite. Then at least one term at the right hand side of (1.28) must be different from zero. This means that at least two different states $\mathbf{k}_r$ and $\mathbf{k}_s$ ($\mathbf{k}_r \neq \mathbf{k}_s$) must have non-vanishing weight ($k_r > 0$, $k_s > 0$). The given state is then said to be divisible. Thus pure states and only these are indivisible. This has been proved in a more exact way by v o n   N e u m a n n [1]).

1.06 *Observation*. In order to establish the observational meaning M_o, one must accept a definite notion of observation. We deal with 3 different notions:

O_c: *the classical notion*: all observables $a(p,q)$ can be measured without fundamental restrictions and without disturbing the system,

O_q: *the quantum notion* (elucidated in **2**): measurement of an observable, which is represented by an operator $\mathbf{a}$, gives as the value of the observable one of the eigenvalues a_μ of $\mathbf{a}$ and leaves the system in the corresponding eigenstate $\mathbf{k}_{\mu\mu}$ (cf. (1.20)); if beforehand the system was in a state $\mathbf{k}$, the probability of this particular measuring result is $Tr(\mathbf{k}\mathbf{k}_{\mu\mu})$.

Suppose for a moment that the statistical description of quantum mechanics had been proven to be formally of the 1st kind S_f^1, but with respect to O_q properly of the 2nd kind S_{oq}^2. Then (if any) the only notion, which could give a proper sense to the formal description, would be

O_u: *the utopian notion*: the uniquely determined processes are observable by methods, hitherto unknown, consistent with and complementary to the methods of O_q.

With respect to quantum theory classical theory is incorrect, though for many purposes it is quite a suitable approximation (for $lim\ \hbar \to 0$). With regard to the utopian conception quantum theory would be correct, but incomplete. In this a description is called correct if none of its statements is in contradiction with observational data. It is called complete if another correct description, which leads to observable statements not contained in the given description, is impossible. This need not imply that all possible observational statements can be derived from a complete theory.

1.07 *The fundamental controversy.* Problem β intends to state certain aspects of the well known controversy about the statistical character of quantum mechanics in a form fit for a reasonable discussion. Such a discussion is only possible as long as the theory is accepted as essentially correct (or rejected and replaced by a more correct theory). The completeness of the theory may be questioned.

The physical reasonings of B o h r a.o. and the mathematical proof of v o n N e u m a n n [1]) (reproduced in 1.08) have shown that (with respect to O_q) the statistical description of quantum mechanics is properly of the 2nd kind S_{oq}^2 (problem β_1). Yet many of the opponents did not throw up the sponge, some because they did not grasp the point, others because they perceived a gap in the reasoning. It seems that a great many of the escapes (as far as they consider quantum mechanics as essentially correct) debouch (if anywhere) into an expectation, which either is already contented with a formal statistical description of the 1ste kind S_f^1, or moreover hopes to give such a description a proper sense of type S_{ou}^1 by proclaiming the utopian notion of observation O_u. The examination of this conception is problem β_2.

Even if one did (we could not satisfactorily) succeed in proving the formal impossibility of type S_f^1 (and consequently of type S_{ou}^1), many of the opponents would not yet strike the flag. We have already gone to meet them in trying to formulize some of their most important objections in a form fit for fruitful discussion. It would be like flogging a dead horse in trying to do so with all vague objections they might possibly raise. Actually that is their own task. If they succeed in doing so, we try to prove the impossibility, they try to find the realization of their (formal or proper) expectations. Formal expectations can be realized by a formal construction, proper ones also require the realization of the type of observations from which they draw their observational meaning. As soon as the opponents succeed in finding a realization, we will (formally or properly) be converted (but not a minute before). As often as we succeed in proving the impossibility, some of the opponents may formulize (if anything) new objections for ever. At best they might be compelled to retreat step by step, they could never be finally vanquished. It may also happen that nobody succeeds in going further. Thus because of running on an infinite track or into a dead one, the controversy may be left undecided. Meanwhile we expect that in an infinite regression the opponents objections will lose more and more interest after every retreat.

1.08 *v o n N e u m a n n's proof*. The only states with a meaning M_{oq} with respect to quantum observations O_q are quantum states (pure states or mixtures). Therefore in a statistical description of the 1st kind S_{oq}^1 a quantum state should be described as a statistical ensemble of quantum states. This is impossible for a pure state, because such a state is indivisible (cf. 1.05). Then the statistical description of quantum mechanics must (with respect to quantum observations) be of the 2nd kind S_{oq}^2. This is in our present mode of expression the point of v o n N e u m a n n's proof [1]). It should be noted that in 1.05 the admission of non-negative probabilities only (non-negative weights and non-negative definite statistical operators) is an essential (and natural) feature of the proof.

Now before going into the details of problem β_2, we first turn to problem α (we need α_5 for β_2).

1.09 *Correspondence $a(p,q) \longleftrightarrow \mathbf{a}$*. In passing from classical to quantum mechanics, the coordinate and momentum q and p, for which

$$(p,q) = 1, \tag{1.29}$$

are replaced by coordinate and momentum operators $\mathbf{q}$ and $\mathbf{p}$, for which

$$[\mathbf{p,q}] = 1 \quad \left(\text{i.e. } \mathbf{pq} - \mathbf{qp} = \frac{\hbar}{i} \right). \tag{1.30}$$

p and q are the generating elements of the commutative ring of classical quantities $a(q,p)$, $\mathbf{p}$ and $\mathbf{q}$ the generating elements of the non-commutative ring of quantum operators $\mathbf{a}$. The non-commutability (1.30) of $\mathbf{p}$ and $\mathbf{q}$ entails that the quantities $a(p,q)$ cannot unambiguously be replaced by $a(\mathbf{p,q})$. The ambiguity is of the order of $\hbar$. The classical quantities $a(p,q)$ can be regarded as approximations to the quantum operators $\mathbf{a}$ for $lim\ \hbar \to 0$. The former can serve as guides to get on the track of the latter. Problem α_1 asks for a rule of correspondence $a(p,q) \to \mathbf{a}$, by which the quantum operators $\mathbf{a}$ can be uniquely determined from the classical quantities $a(p,q)$.

In practical problems no fundamental difficulties seem to occur in finding the appropriate form of the required operators **a**. This suggests the problem (not further discussed here) whether all or only a certain simple class of operators **a** occur in quantum mechanics.

Suppose for a moment that all relevant quantum operators **a** had been fixed in one or other way. Then one might ask for a rule **a** → $a(p,q)$, by which the corresponding classical quantities $a(p, q)$ are uniquely determined (problem α_2). Problem α_2 would be easily solved in zero order of $\hbar$, ambiguities might arise in higher order. Now (with respect to O_q) the classical quantities have only a meaning as approximations to the quantum operators for $lim\ \hbar \to 0$. Therefore, whereas in zero order of $\hbar$ it is hardly a problem, in higher order problem α_2 has no observational meaning M_{oq} (with respect to O_q).

Problems α_1 and α_2 could be combined into problem α_3, asking for a rule of one-to-one correspondence $a(p,q)$ ←→ **a** between the classical quantities $a(p,q)$ and the quantum operators **a**. Beyond the trivial zero order stage in $\hbar$, problem α_3 can (with respect to O_q) only have an observational meaning M_{oq} as a guiding principle for detecting the appropriate form of the quantum operators (i.e. as problem α_1). A formal solution of problem α_3 has been proposed by W e y l [2]) (cf. 4.03). We incidentally come back to problem α_3 in 1.18.

1.10 *Quantum observables.* In this section a will not denote a classical quantity $a(p,q)$, but it will stand as a symbol for the observable, which (with regard to O_q) is represented by the quantum operator **a**. According to O_q two or more observables $a, b, \ldots$ can be simultaneously measured or not, according as the corresponding operators **a**, **b**, $\ldots$ respectively do or do not commute i.e. as they have all eigenstates in common or not. Problem α_4 deals with the (one-to-one) correspondence a ←→ **a** between the symbols a and the operators **a**. Problem α_4 has no sense as long as the symbols a are undefined. They may, however, be implicitely defined just by putting a rule of correspondence. (When the symbols a are identified with the classical quantities $a(p,q)$, problem α_4 becomes identical with problem α_3). V o n N e u m a n n [1]) has proposed the rules

$$\text{if } a \longleftrightarrow \mathbf{a}, \text{ then } f(a) \longleftrightarrow f(\mathbf{a}), \qquad\qquad \text{I}$$

$$\text{if } a \longleftrightarrow \mathbf{a} \text{ and } b \longleftrightarrow \mathbf{b}, \text{ then } a + b \longleftrightarrow \mathbf{a} + \mathbf{b}. \qquad \text{II}$$

$f(\mathbf{a})$ is defined as the operator, which has the same eigenstates as **a** with eigenvalues $f(a_\mu)$, where a_μ are those of **a**. Then I seems to be obvious. The observable $f(a)$ can be measured simultaneously with

a, its value is $f(a_\mu)$, where a_μ is that of a. When **a** and **b** commute, **a** + **b** has the same eigenstates as **a** and **b** with eigenvalues $a_\mu + b_\mu$, where a_μ and b_μ are those of **a** and **b**. Then II seems also to be obvious. $a + b$ can be measured simultaneously with a and b, its value is $a_\mu + b_\mu$, where a_μ and b_μ are the values of a and b. When **a** and **b** do not commute, II is proposed with some hesitation. Because according to O_q the probability of finding a value a_μ for a in a state **k** is $Tr(\mathbf{kk}_{\mu\mu})$ (and because of 1.20)), the expectation value (average value) of a in this state is

$$Ex(a) = \sum_\mu Tr(\mathbf{kk}_{\mu\mu})a_\mu = Tr(\mathbf{ka}) \qquad (1.31)$$

and similar for b. If one requires that for a certain pair of observables a and b always

$$Ex(a + b) = Ex(a) + Ex(b), \qquad (1.32)$$

one must, because of

$$Tr(\mathbf{k}(\mathbf{a} + \mathbf{b})) = Tr(\mathbf{ka}) + Tr(\mathbf{kb}), \qquad (1.33)$$

have that

$$Ex(a + b) = Tr(\mathbf{k}(\mathbf{a} + \mathbf{b})). \qquad (1.34)$$

Because this has to hold for all states **k**, a and b have to satisfy rule II. When II is given up for certain pairs a,b, the additivity of the expectation values of these pairs has also to be given up.

In 4.01 it will be shown that, if I and II shall be generally valid, the symbols a have to be isomorphic with the operators **a**. But then there is no reason to introduce the former, their task (if any) can be left to the latter. Accordingly for the sake of brevity we shall henceforth speak of the (quantum) observable **a**.

When on the other hand, the symbols a are intended as real commuting quantities, the general validity of I and II cannot be maintained. As long as the symbols a are not further defined, problem α_4 comes to searching for a one-to-one correspondence $a \longleftrightarrow$ **a** between the commutative ring of real symbols a and the non-commutative ring of H e r m i t i a n operators **a**. There may be no, one or more solutions. After the pleas for I and for II, one might be inclined to maintain 1 and to restrict II. In 1.13 we meet with a particular case (problem α_5) for which II has to be maintained and therefore I has to be restricted. Because we are further exclusively interested in problem α_5, we will not examine the possibility of solutions for which II is restricted.

1.11 *Hidden parameters.* We try to trace the conditions for the assumption that the statistical description of quantum mechanics is (at least formally) of the 1st kind S^1 (problem β). A statistical description S^1 must be obtained by statistical averaging over uniquely determined processes. The averaging must be described by integration or summation over a statistical distribution with respect to certain parameters. Unless they are further specified, we denote all parameters by a single symbol ξ and integration (including a possible density function) and summation over continuous and discrete parameters by $\int d\xi$. Parameters, which are in no way observable with respect to O_q, are called hidden parameters. (We exclude their occurence in 1.15). As a pure superstate we define a state for which all parameters (inclusive the hidden ones) have a definite value.

1.12 *Distributions.* A quantum state must be described as an ensemble of pure superstates. The statistical operator **k** of the quantum state must correspond to at least one (non-negative definite) distribution function $k(\xi)$ for the superstates. For each definite value of ξ all $k(\xi)$ must have definite values and therefore must commute. $k(\xi)$ must be normalized by $\int d\xi\, k(\xi) = 1$, so that with (1.23)

$$Tr\mathbf{k} = \int d\xi\, k(\xi). \tag{1.35}$$

Further the correspondence must be linear

$$\text{if } \mathbf{k}_1 \longleftrightarrow k_1(\xi) \text{ and } \mathbf{k}_2 \longleftrightarrow k_2(\xi), \text{ then } \mathbf{k}_1 + \mathbf{k}_2 \longleftrightarrow k_1(\xi) + k_2(\xi). \tag{1.36}$$

The observable (with respect to O_q) represented by the statistical operator **k** of a pure quantum state has the eigenvalue 1 in this quantum state and 0 in all orthogonal states. The probability of measuring in a system, which is originally in a quantum state **k**, the value 1 (and leaving the system in the pure quantum state $\mathbf{k}_{\mu\mu}$) is $Tr(\mathbf{k}\mathbf{k}_{\mu\mu})$. In a description of type S^1 this probability must be interpreted as the probability that any superstate belonging to the ensemble with distribution function $k(\xi)$ corresponding to **k** also belongs to the ensemble with distribution function $k_{\mu\mu}(\xi)$ corresponding to $\mathbf{k}_{\mu\mu}$. The latter probability is $\int d\xi\, k(\xi)k_{\mu\mu}(\xi)$. Therefore the correspondence $\mathbf{k} \longleftrightarrow k(\xi)$ must be so that always

$$Tr(\mathbf{k}_1\mathbf{k}_2) = \int d\xi\, k_1(\xi)k_2(\xi). \tag{1.37}$$

For two orthogonal states k_1 and k_2 this expression is zero, which guarantees that the distribution functions $k_1(\xi)$ and $k_2(\xi)$ do not overlap, provided they are non-negative definite.

1.13 *Superquantities.* The expectation value of the observable **a** in the quantum state **k** is because of (1.31) and (1.37)

$$\sum_{\mu} Tr(\mathbf{k}\mathbf{k}_{\mu\mu})a_{\mu} = \sum_{\mu} \int d\xi \, k(\xi) k_{\mu\mu}(\xi) a_{\mu}. \qquad (1.38)$$

The right hand member of (1.38) can be interpreted as the average value of a quantity $a(\xi) = \sum_{\mu} a_{\mu} k_{\mu\mu}(\xi)$ (defined as the superquantity corresponding to the observable **a**) in the ensemble of superstates with distribution function $k(\xi)$. This is exactly the way in which the expectation value should appear in a description of type S^1. Thus with the correspondence **a** $\longleftrightarrow a(\xi)$ (which is a linear generalization of **k** $\longleftrightarrow k(\xi)$) the expectation value of **a** in the state **k** can be written

$$Tr(\mathbf{k}\mathbf{a}) = \int d\xi \, k(\xi) \, a(\xi). \qquad (1.39)$$

Comparison with (1.35) shows that the unit operator **1** has to correspond to the unit quantity 1

$$\mathbf{1} \longleftrightarrow 1. \qquad \qquad \text{III}$$

By a further linear generalization of (1.39) we see that the correspondence **a** $\longleftrightarrow a(\xi)$ must obey the rule

if **a** $\longleftrightarrow a(\xi)$ and **b** $\longleftrightarrow b(\xi)$, then $Tr(\mathbf{ab}) = \int d\xi \, a(\xi) \, b(\xi)$.   IV

Rule II is a consequence of rule IV (the necessity of II is evident from the beginning, because average of sum = sum of averages). Therefore rule I cannot be satisfied without restrictions.

Problem α_5 is how to establish the correspondence **a** $\longleftrightarrow a(\xi)$. α_5 is, like α_3, a special case of α_4.

1.14 *Equations of motion.* The equations of motion for the quantum states must be obtained from the equations of motion for the superstates. The former are determined by the H a m i l t o n i a n operator **H** (which may depend on time t) of the system according to the equation of motion of the statistical operator **k**

$$\frac{d\mathbf{k}}{dt} = - [\mathbf{H},\mathbf{k}] \qquad (1.40)$$

(which is equivalent to the S c h r ö d i n g e r equation

$$-\frac{\hbar}{i} \frac{\partial \varphi}{\partial t} = \mathbf{H}\varphi$$

for pure quantum states). Because the correspondence $\mathbf{k} \longleftrightarrow k(\xi)$ is linear, we have

$$\frac{d\mathbf{k}}{dt} \longleftrightarrow \frac{dk(\xi)}{dt} . \tag{1.41}$$

(1.40) can be integrated into

$$\mathbf{k}(t) = e^{-\frac{i}{\hbar}\int_{t_0}^{t} dt' \, \mathbf{H}(t')} \, \mathbf{k}(t_0) \, e^{\frac{i}{\hbar}\int_{t_0}^{t} dt' \, \mathbf{H}(t')} \tag{1.42}$$

(which is equivalent to $\varphi(t) = e^{-\frac{i}{\hbar}\int_{t_0}^{t} dt' \, \mathbf{H}(dt')} \varphi(t_0)$ for pure quantum states). If the superquantity corresponding to the bracket expression $[\mathbf{a},\mathbf{b}]$ is written $((a(\xi), b(\xi)))$ (the former and consequently also the latter bracket expression is antisymmetrical), the equation of motion of the distribution function $k(\xi)$ reads

$$\frac{dk(\xi)}{dt} = - ((H(\xi), k(\xi))). \tag{1.43}$$

Because

$$\frac{d}{dt} Tr(\mathbf{k}\mathbf{a}) = Tr\left(- [\mathbf{H},\mathbf{k}] \, \mathbf{a} + \mathbf{k}\frac{\partial \mathbf{a}}{\partial t}\right) = Tr\left(\mathbf{k}\left([\mathbf{H},\mathbf{a}] + \frac{\partial \mathbf{a}}{\partial t}\right)\right) \tag{1.44}$$

and correspondingly

$$\frac{d}{dt} \int d\xi \, k(\xi) \, a(\xi) = \int d\xi \left(- ((H(\xi), k(\xi))) \, a(\xi) + k(\xi) \, \frac{\partial a(\xi)}{\partial t}\right)$$
$$= \int d\xi \, k(\xi) \left(((H(\xi), a(\xi))) + \frac{\partial a(\xi)}{\partial t}\right), \tag{1.45}$$

the dynamical time dependence can be shifted from the wave functions φ and the statistical operators $\mathbf{k}$ (S c h r ö d i n g e r representation) and the distribution functions $k(\xi)$ to the operators $\mathbf{a}$ (H e i s e n b e r g representation) and the superquantities $a(\xi)$.
Instead of (1.40), (1.43) we then get

$$\frac{d\mathbf{a}}{dt} = \frac{\partial \mathbf{a}}{\partial t} + [\mathbf{H}, \mathbf{a}], \tag{1.46}$$

$$\frac{da(\xi)}{dt} = \frac{\partial a(\xi)}{\partial t} + ((H(\xi), a(\xi))). \tag{1.47}$$

For those parameters ξ, which correspond to observable quantities (with respect to O_q) (1.47) must be valid and reads

$$\frac{d\xi}{dt} = \frac{\partial \xi}{\partial t} + ((H(\xi), \xi)). \tag{1.48}$$

The equations of motion for the hidden parameters may be of a different form. When all parameters (inclusive the hidden ones) are continuous, their equations of motion have to satisfy the condition that when inserted in

$$\frac{da(\xi)}{dt} = \frac{\partial a(\xi)}{\partial t} + \frac{\partial a(\xi)}{\partial \xi} \frac{d\xi}{dt} \qquad (1.49)$$

(where the last term stands symbolically for a sum over all separate parameters ξ), they must give (1.47).

We may summarize that, in order to give a statistical description of the 1st kind, one would have to determine (only formally for type S_f^1, also experimentally for type S_o^1) the parameters ξ (inclusive the hidden ones) and the density function, the (one-to-one or one-to-many) correspondence $\mathbf{a} \longleftrightarrow a(\xi)$ (problem α_5) and the equations of motion for the hidden parameters (if there are any such), all with regard to the imposed conditions.

1.15 *Correspondence* $\mathbf{a} \longleftrightarrow a(\xi)$. Because a non-Hermitian operator $\mathbf{a}$ (with adjoint $\mathbf{a}^\dagger$) can be written as a complex linear combination of Hermitian operators

$$\mathbf{a} = \tfrac{1}{2} (\mathbf{a} + \mathbf{a}^\dagger) + \frac{1}{2i} (i\mathbf{a} - i\mathbf{a}^\dagger),$$

the generalization of the correspondence $\mathbf{a} \longleftrightarrow a(\xi)$ to non-Hermitian operators is uniquely determined. Now take the non-Hermitian transition operators $\mathbf{k}_{\mu\nu}$, which according to (1.13), (1.14) form a complete orthonormal system in the ring of operators $\mathbf{a}$. For the corresponding functions $k_{\mu\nu}(\xi)$ we get corresponding to (1.11), (1.12); (1.13), (1.14) and (1.15) (and using III, IV and (1.03)) the relations

$$\int d\xi \, k_{\mu\nu}(\xi) = \delta_{\mu\nu}, \qquad (1.50)$$

$$\sum_{\mu} k_{\mu\mu}(\xi) = 1; \qquad (1.51)$$

$$\int d\xi \, k^*_{\mu\nu}(\xi) \, k_{\mu'\nu'}(\xi) = \delta_{\mu\mu'} \, \delta_{\nu\nu'}, \qquad (1.52)$$

$$\sum_{\mu,\nu} k_{\mu\nu}(\xi) \, k^*_{\mu\nu}(\xi') = \delta(\xi - \xi') \qquad (1.53)$$

($\delta(\xi - \xi')$ stands for a product of δ-symbols for all parameters ξ and the inverse of the density function) and

$$a(\xi) = \sum_{\mu,\nu} \alpha_{\nu\mu} k_{\mu\nu}(\xi) \quad \text{with} \quad \alpha_{\nu\mu} = \int d\xi \, k^*_{\mu\nu}(\xi) \, a(\xi). \qquad (1.54)$$

(the $\alpha_{\nu\mu}$ are the same as in (1.15)). These relations show, that the functions $a(\xi)$ can be regarded as elements of a (gereralized) H i l b e r t space, in which the $k_{\mu\nu}(\xi)$ form a complete orthonormal system; (1.52) expresses the orthonormality, (1.53) the completeness.

We now show that the correspondence $\mathbf{a} \longleftrightarrow a(\xi)$ has to be a one-to-one correspondence. Suppose for a moment there are operators $\mathbf{k}_{\mu\nu}$ to which there correspond more than one functions $k_{\mu\nu}(\xi)$, which we distinguish by an index ρ, $\mathbf{k}_{\mu\nu} \longleftrightarrow k_{\mu\nu;\rho}(\xi)$. Then the expression

$$\sum_{\mu',\nu'} \int d\xi\, k_{\mu\nu;\rho}(\xi)\, k^{*}_{\mu'\nu';\rho'}(\xi)\, k_{\mu'\nu';\rho''}(\xi')$$

evaluated with (1.52) gives $k_{\mu\nu;\rho''}(\xi')$, evaluated with (1.53) it gives $k_{\mu\nu;\rho}(\xi')$. Therefore $k_{\mu\nu;\rho''}(\xi')$ and $k_{\mu\nu;\rho}(\xi')$ have to be identical. To each operator $\mathbf{a}$ and only to this one there has to correspond one and only one superquantity $a(\xi)$. As a consequence the superquantities $a(\xi)$ must depend on the same number of parameters (at least if they are not too bizarre) as the operators $\mathbf{a}$, i.e. on twice as many as the wave functions φ.

Thus to each (normalizable) real function $a(\xi)$ and only to this one there corresponds one and only one H e r m i t i a n operator $\mathbf{a}$, which represents an observable quantity (with respect to O_q). In other words every real function $a(\xi)$ is a superquantity. Because this also holds for the (real and imaginary parts of the) parameters ξ themselves, none of them can be hidden in the sense defined above. (An observable quantity may occasionally be inobservable in a measuring device adepted to an incommensurable quantity; in this sense a parameter may occasionally be hidden). In particular all parameters must obey (1.48).

Comparing (1.15) and (1.54) we see that the correspondence $\mathbf{a} \longleftrightarrow a(\xi)$ can be expressed by

$$a(\xi) = Tr(\mathbf{m}(\xi)\mathbf{a}), \quad \mathbf{a} = \int d\xi\, \mathbf{m}(\xi)a(\xi), \tag{1.55}$$

with

$$\mathbf{m}(\xi) = \sum_{\mu,\nu} \mathbf{k}_{\mu\nu}\, k^{*}_{\mu\nu}(\xi); \quad \mathbf{m}^{\dagger}(\xi) = \mathbf{m}(\xi). \tag{1.56}$$

The H e r m i t i a n transformation nucleus $\mathbf{m}(\xi)$ satisfies the relations

$$Tr\,\mathbf{m}(\xi) = 1, \tag{1.57}$$

$$\int d\xi\, \mathbf{m}(\xi) = \mathbf{1}; \tag{1.58}$$

$$Tr(\mathbf{m}(\xi)\, \mathbf{m}(\xi')) = \delta(\xi - \xi'), \tag{1.59}$$

$$\int d\xi\, Tr(\mathbf{m}(\xi)\, \mathbf{a})\, Tr(\mathbf{m}(\xi)\, \mathbf{b}) = Tr(\mathbf{ab}) \text{ (for every } \mathbf{a} \text{ and } \mathbf{b}) \tag{1.60}$$

(1.60) is equivalent to

$$\int d\xi \; \mathbf{m}(\xi) \, Tr(\mathbf{m}(\xi) \, \mathbf{a}) = \mathbf{a} \text{ (for every } \mathbf{a}).$$

(1.59) expresses that $\mathbf{m}(\xi)$ is orthonormal with respect to the ring of operators $\mathbf{a}$, complete with respect to the ring of superquantities $a(\xi)$; (1.60) expresses the crossed properties.

If, on the other hand, a H e r m i t i a n transformation nucleus $\mathbf{m}(\xi)$ satisfies the conditions (1.57), (1.58); (1.59), (1.60), the correspondence (1.55) satisfies III and IV. We may either choose a complete orthonormal system of $\mathbf{k}_{\mu\nu}$, satisfying (1.11), (1.12); (1.13), (1.14) and determine the corresponding system of $k_{\mu\nu}(\xi)$, which then satisfy (1.50), (1.51); (1.52), (1.53), or we choose the latter system and determine the former one. In both cases $\mathbf{m}(\xi)$ can be expanded according to (1.56).

1.16 *Uniqueness.* Now let us see whether the correspondence $\mathbf{a} \longleftrightarrow a(\xi)$ is uniquely determined by the conditions III and IV. Suppose we have two different nuclei $\mathbf{m}'(\xi)$ and $\mathbf{m}''(\xi)$, depending on the same parameter ξ and both satisfying (1.57), (1.58); (1.59), (1.60). When we choose for both the same complete orthonormal system of $k_{\mu\nu}(\xi)$ satisfying (1.50), (1.51); (1.52), (1.53), we find two corresponding systems of $\mathbf{k}'_{\mu\nu}$ and $\mathbf{k}''_{\mu\nu}$, which each satisfy (1.11), (1.12); (1.13), (1.14). Therefore the latter systems can be connected by a unitary transformation

$$\mathbf{k}'_{\mu\nu} = \mathbf{u}\mathbf{k}''_{\mu\nu}\,\mathbf{u}^\dagger, \; \mathbf{u}\mathbf{u}^\dagger = 1; \; \mathbf{k}''_{\mu\nu} = \mathbf{u}^\dagger\mathbf{k}'_{\mu\nu}\,\mathbf{u} \qquad (1.61)$$

(expressed analoguous to (1.03) $\mathbf{u}$ can be written as $\sum\limits_{\mu} \varphi'_\mu \, \varphi''^\dagger_\mu$). The same unitary transformation connects the nuclei $\mathbf{m}'(\xi)$ and $\mathbf{m}''(\xi)$ and also the statistical operators $\mathbf{k}'$ and $\mathbf{k}''$ corresponding to the same distribution function $k(\xi)$ and the operators $\mathbf{a}'$ and $\mathbf{a}''$ corresponding to the same superquantity $a(\xi)$. Then the single and double dashed representations are isomorphous and in quantum mechanics regarded as equivalent. Therefore, when the parameters ξ have been chosen, the correspondence $\mathbf{a} \longleftrightarrow a(\xi)$ (if there is any correspondence) can be considered as unique.

When we choose one set of parameters ξ and another set of parameters η, the nuclei $\mathbf{m}(\xi)$ and $\mathbf{m}(\eta)$ (if there are any nuclei) can be considered as uniquely determined. When we take a complete orthonormal system of $\mathbf{k}_{\mu\nu}$ satisfying (1.11), (1.12); (1.13), (1.14), we find two corresponding systems of $k'_{\mu\nu}(\xi)$ and $k''_{\mu\nu}(\eta)$, which each satisfy

(1.50), (1.51); (1.52), (1.53). Then it follows that the superquantities $a'(\xi)$ and $a''(\eta)$, corresponding to the same operator **a** are connected by

$$a'(\xi) = \int d\eta\, v(\xi; \eta) a''(\eta); \quad a''(\eta) = \int d\xi\, v(\xi; \eta) a'(\xi), \quad (1.62)$$

where the transformation nucleus

$$v(\xi; \eta) = \sum_{\mu,\nu} k'_{\mu\nu}(\xi)\, k''^{*}_{\mu\nu}(\eta); \quad v(\xi; \eta) = v^{*}(\xi; \eta) \quad (1.63)$$

satisfies

$$\int d\xi\, v(\xi; \eta) = \int d\eta\, v(\xi; \eta) = 1; \quad (1.64)$$

$$\int d\eta\, v(\xi; \eta)\, v(\xi'; \eta) = \delta(\xi - \xi'), \quad \int d\xi\, v(\xi; \eta)\, v(\xi; \eta') = \delta(\eta - \eta'). \quad (1.65)$$

The rings of $a'(\xi)$ and of $a''(\eta)$ are not necessarily isomorphous. When they are, we must have

$$\iint d\eta' d\eta''\, v(\xi; \eta')\, v(\xi; \eta'')\, a''(\eta')\, b''(\eta'') = \int d\eta\, v(\xi; \eta)\, a''(\eta)\, b''(\eta) \quad (1.66)$$

for every $a''(\eta)$ and $b''(\eta)$, which requires

$$v(\xi; \eta')\, v(\xi; \eta'') = v(\xi; \eta')\, \delta(\eta' - \eta'') \quad (1.67)$$

and similarly

$$v(\xi'; \eta)\, v(\xi''; \eta) = v(\xi'; \eta)\, \delta(\xi' - \xi''). \quad (1.68)$$

The solutions of (1.67) and (1.68) have the form

$$v(\xi; \eta') = \delta(\eta(\xi) - \eta') \quad (1.69)$$

and

$$v(\xi'; \eta) = \delta(\xi' - \xi(\eta)), \quad (1.70)$$

where $\eta(\xi)$ and $\xi(\eta)$ are single valued functions. Because (1.69) and (1.70) have to be identical, $\eta(\xi)$ and $\xi(\eta)$ have to be inverse to each other with unit functional determinant

$$\left| \frac{\partial(\eta)}{\partial(\xi)} \right| = \left| \frac{\partial(\xi)}{\partial(\eta)} \right| = 1 \quad (1.71)$$

(it should be remembered that we symbolically write ξ or η for what might be a whole set of parameters ξ or η). With (1.69), (1.70) we get for (1.62)

$$a'(\xi) = a''(\eta(\xi)); \quad a''(\eta) = a'(\eta(\xi)). \quad (1.72)$$

This shows that the transformation between two isomorphous representations $a'(\xi)$ and $a''(\eta)$ can be regarded as merely a transformation of the parameters. It further follows that, if the dynamical conditions for (1.49) are fulfilled by one of these representations, they are also fulfilled by the other one. Therefore isomorphous representations can be regarded as equivalent.

When the solution $v(\xi;\eta)$ of (1.64), (1.65) is not of the form (1.69), (1.70), the representations $a'(\xi)$ and $a''(\eta)$ are non-isomorphous.

1.17 *Parameters*. In 4.03 we derive a correspondence, satisfying III and IV, in which the two independent parameters (denoted by p and q), which run continuously between $-\infty$ and $+\infty$, correspond to the operators $\mathbf{p}$ and $\mathbf{q}$. This choice of parameters might seem the most satisfactory one, as it is adapted to the fundamental part played by the momentum and the coordinate. (By the way, because momentum and coordinate cannot simultaneously be measured, p may be regarded as occasionally hidden in a coordinate measurement, q similarly in a momentum measurement — or in a somewhat different conception p may be regarded as occasionally hidden in q-representation, q in p-representation; both p and q may be regarded as occasionally partially hidden in other measurements or representations).

In 1.16 we have seen that for each choice of a complete orthonormal system of $k_{\mu\nu}(p,q)$, satisfying (1.50), (1.51); (1.52), (1.53), there must for every other representation with parameters ξ be a similar system of $k_{\mu\nu}(\xi)$ with the same set of indices μ,ν. That makes us expect that when ξ stands for a set of not too bizarre continuous parameters, the latter can like p and q be represented by two independent real parameters r and s. We do not examine the validity of this expectation (which would be very difficult).

1.18 *Bracket expressions*. When these parameters r and s are also independent of time, the consistency relation for (1.47), (1.48) and (1.49) reads

$$((H(r,s), a(r,s))) = \frac{\partial a(r,s)}{\partial r} ((H(r,s), r)) + \frac{\partial a(r,s)}{\partial s} ((H(r,s), s)) \quad (1.73)$$

$$\text{(for every } a(r,s)).$$

When the superquantities $H(r,s)$ corresponding to the H a m i l t o n i a n operators $\mathbf{H}$ are not restricted to functions of a too special type, (1.73) requires (using the antisymmetry properties

$$((r,s)) = -((s,r)); ((r,r)) = ((s,s)) = 0)$$

$$((a(r,s), b(r,s))) = ((r,s))(a(r,s), b(r,s)) \text{ (for every } a(r,s) \text{ and } b(r,s)), \quad (1.74)$$

with the P o i s s o n brackets (similar to (1.01))

$$(a(r,s), b(r,s)) = \frac{\partial a(r,s)}{\partial r}\frac{\partial b(r,s)}{\partial s} - \frac{\partial a(r,s)}{\partial s}\frac{\partial b(r,s)}{\partial r}. \quad (1.75)$$

For the superquantities $p(r,s)$ and $q(r,s)$ corresponding to **p** and **q** we get because of (1.30)

$$((p(r,s), q(r,s))) = ((r,s)) (p(r,s), q(r,s)) = 1. \qquad (1.76)$$

Therefore (1.74) can also be written .

$$((a(r,s), b(r,s))) = \frac{(a(r,s), b(r,s))}{(p(r,s), q(r,s))}. \qquad (1.77)$$

This means that the correspondence **a** $\longleftrightarrow a(r,s)$ has to satisfy the rule

if   **a** $\longleftrightarrow a(r,s)$, **b** $\longleftrightarrow b(r,s)$ and **p** $\longleftrightarrow p(r,s)$, **q** $\longleftrightarrow q(r,s)$,

$$\text{then} \quad [\mathbf{a},\mathbf{b}] \longleftrightarrow \frac{(a(r,s), b(r,s))}{(p(r,s), q(r,s))}. \qquad \text{V}$$

The analoguous derivation for the parameters p and q gives (independent of our unproved expectation about the parameters r and s) the condition

if **a** $\longleftrightarrow a(p,q)$ and **b** $\longleftrightarrow b(p,q)$, then [**a**,**b**] $\longleftrightarrow (a(p,q), b(p,q))$. V′

For this choice of parameters problem α_5 of the correspondence between the superquantities $a(p,q)$ and the quantum operators **a** seems very similar to problem α_3 of the correspondence between the classical quantities $a(p,q)$ and the quantum operators **a**, by which they are replaced in the procedure of quantization. The fact that in this procedure the P o i s s o n brackets in the equations of motion are replaced by operator brackets might suggest rule V′ in problem α_3. If a solution of α_3 satisfying rules III, IV and V′ could be found, the classical description could be regarded as the description of the uniquely determined processes in a statistical description of the 1st kind S^1. The utopian notion O_u, intended to proclaim these processes as observable, would coincide with the classical notion O_c. This would not (as it might seem) exactly mean a return towards the old classical theory, which was regarded as incorrect (with respect to O_q and therefore also with respect to O_u, which regards O_q as correct, though incomplete), because one would have to deal with peculiar distributions of classical systems. These distributions would have to be responsible for quantization.

But such a solution cannot be found. In 4.02 we show that V′ is self contradictory (except for $lim\ \hbar \rightarrow 0$). Because V′ already fails

for operators of occuring types, a restriction of the admitted operators could not help. Therefore a solution of problem α_5 with p and q
as parameters, which satisfies the dynamical conditions, is impossible, just as a solution of α_3, which describes the quantization of the
classical equations of motion by the same rule as the quantization of
the classical observables.

This is in point of fact all we have been able to prove. Though p
and q may seem the most satisfactory choice of parameters in a
description of type S^1, the formal disproof of just this description
does not involve the impossibility of any description of type S_f^1. A
complete proof of the impossibility of a description of type S_f^1 does not
seem simple and neither does the construction of such a description.

For a pair of continuous time independent parameters r and s
condition V would have to be satisfied. When the commutator of
r and s commutes with r and s, V is self contradictory just like V′.
It is doubtful whether V can be consistent in other cases. A pair of
continuous time dependent parameters $r(t)$ and $s(t)$ must at every
time t be unique single-valued functions of the initial values $r(t_0)$ and
$s(t_0)$ at an arbitrary time t_0. Then instead of the time dependent $r(t)$
and $s(t)$ the time independent $r(t_0)$ and $s(t_0)$ can serve as parameters.
Therefore, if our expectation about continuous parameters is justified, the difficulty for such parameters lies mainly in the consistency
of V. It is difficult to see how parameters with entirely or partially
discrete values or of too bizarre continuous type could give a satisfactory description of type S^1.

There are still more difficulties for a description S^1 as we will see
in a moment.

1.19 *Quasi-statistical description*. Whereas it is doubtful whether
the dynamical condition V can be fulfilled, conditions III and IV can
be satisfied without much difficulties. With a solution of the latter
conditions only, one can construct a quasi-statistical description of
the 1st kind Q^1, which looks very similar to a formal statistical
description of the 1st kind S_f^1, but in general does not satisfy the
dynamical (and, as we will see in a moment, other necessary) conditions. A solution of III and IV gives according to (1.39) the correct
average values. But the real distribution function $k(\xi)$ corresponding
to a H e r m i t i a n non-negative definite statistical operator **k** of a
quantum state (pure state or mixture) is in general not non-negative
definite.

The difficulty of interpreting negative probabilities might perhaps be surmountable, at least in formal sense M_f. Meanwhile, according to the remark following (1.37), it is no longer guaranteed, that the distribution functions $k_1(\xi)$ and $k_2(\xi)$ corresponding to orthogonal quantum states $\mathbf{k}_1$ and $\mathbf{k}_2$ do not overlap. And overlapping of such distribution functions it not allowed by the notion of quantum observability O_q. We see this in the following way. Suppose we subject the system repeatedly to a measurement, which distinguishes between the states k_1 and k_2 (and other orthogonal states). When after one measurement the system is left in the state k_1, the probability of finding it after a repeated measurement in the state k_2 is 0 because of (1.37). In the quantum mechanical interpretation this means absolute certainty of not finding the state k_2. In the quasi-statistical interpretation the zero value for the right hand member of (1.37) results from integration of positive and negative probabilities over the region of overlapping. Integration over a statistical distribution refers to a great number of measurements. In a proper statistical description of the 1st kind S^1 the absolute certainty of not finding the state k_2, even in a single measurement, can only be established if no superstate occurring in the ensemble $k_1(\xi)$ can also occur in the ensemble $k_2(\xi)$, i.e. if $k_1(\xi)$ and $k_2(\xi)$ do not overlap.

Therefore in order to find a statistical description of type S_f^1, one would have to satisfy not only conditions II, IV and V (or another dynamical condition), but also the condition that the distribution functions of quantum states are non-negative definite, or at least that the distribution functions of orthogonal states do not overlap. This task does not look very promising.

We incidentally remark that in any representation of type Q^1 either of the two parameters can be treated as occasionally hidden. Already after integration over this one parameter we get the quantum mechanical formalism in the representation of the other parameter. In particular no negative probabilities are left.

In 4.03 we derive a particular solution (W e y l's correspondence) of III and IV with parameters p and q and in **5** we discuss the quasi-statistical description Q^1 to which it leads. We do so not only for the sake of curiosity, but also because it is very instructive to those opponents in the fundamental controversy, who have a description of type S^1 (similar to that of classical statistical mechanics) vaguely in mind. A description of type Q^1 might be the utmost (though rather poor) realization of such foggy ideas. (The mysterious hidden parameters then turn out as ordinary, occasionally inobservable, observables). Such a description clearly shows the obstacles (equations of motion; non-negative probabilities or non-overlapping distributions) at which all such conceptions may be expected to break down.

So far the general line of reasoning. Before dealing further with correspondence in **4**, for which we need the operator relations of **3**, we review in **2** the measuring process in terms of the operator representation.

2. The measuring process.

2.01 *Deviation.* Quite apart from the interpretation of 1.10, the expectation value of a quantum observable **a** in a quantum state **k** is given by (1.31) or

$$Ex(\mathbf{k};\mathbf{a}) = Tr(\mathbf{ka}). \tag{2.01}$$

Further the deviation of this observable in this state is defined by

$$Dev(\mathbf{k};\mathbf{a}) = Ex\big(\mathbf{k}; (\mathbf{a} - \mathbf{1}Ex(\mathbf{k};\mathbf{a}))^2\big) = Tr\big(\mathbf{k}(\mathbf{a} - \mathbf{1}Tr(\mathbf{ka}))^2\big) =$$
$$= Tr(\mathbf{ka}^2) - (Tr(\mathbf{ka}))^2. \tag{2.02}$$

First we review some consequences of this definition, detached of any interpretation.

It can be seen from the inner members of (2.02) that the deviation is non-negative. We form the transition operators $\mathbf{k}_{\nu\mu}$ (1.03) of the complete system of eigenfunctions φ_μ of **a** with eigenvalues a_μ and expand **k** according to (1.15) as

$$\mathbf{k} = \sum_{\mu,\nu} \varkappa_{\nu\mu}\, \mathbf{k}_{\mu\nu} \text{ with } \varkappa_{\nu\mu} = Tr(\mathbf{k}_{\nu\mu}\,\mathbf{k}). \tag{2.03}$$

The normalization of **k** $(Tr\mathbf{k} = 1)$ gives with (1.11)

$$\sum_{\mu} \varkappa_{\mu\mu} = 1. \tag{2.04}$$

Then (2.02) gives

$$Dev(\mathbf{k};\mathbf{a}) = \sum_{\mu} \varkappa_{\mu\mu}\, a_\mu^2 - \Big(\sum_{\mu} \varkappa_{\mu\mu}\, a_\mu\Big)^2 = \tfrac{1}{2}\sum_{\mu,\nu} \varkappa_{\mu\mu}\varkappa_{\nu\nu}\,(a_\mu - a_\nu)^2. \tag{2.05}$$

If **k** is a pure state with wave function φ, we have

$$\varkappa_{\mu\mu} = Tr(\mathbf{k}_{\mu\mu}\,\mathbf{k}) = |\varphi_\mu^\dagger \varphi|^2. \tag{2.06}$$

$\varkappa_{\mu\mu}$ is then non-negative and (2.05) can only be zero, if φ is a linear combination of eigenfunctions φ_μ all with the same eigenvalue a_μ.

If the normalized quantum state **k** (pure state or mixture) can be written as a mixture of other normalized states $\mathbf{k}_r$ with weights k_r

$$\mathbf{k} = \Sigma_r k_r \mathbf{k}_r; \quad k_r \geqslant 0, \quad \Sigma_r k_r = 1, \tag{2.07}$$

(2.02) gives

$$Dev(\mathbf{k};\mathbf{a}) = \Sigma_r k_r Tr(\mathbf{k}_r\mathbf{a}^2) - (\Sigma_r k_r Tr(\mathbf{k}_r\mathbf{a}))^2$$

$$= \Sigma_r k_r Dev(\mathbf{k}_r;\mathbf{a}) + \tfrac{1}{2} \Sigma_{r,s} k_r k_s (Ex(\mathbf{k}_r;\mathbf{a}) - Ex(\mathbf{k}_s;\mathbf{a}))^2. \tag{2.08}$$

The deviation of **a** in the state **k** is therefore only zero, if all occuring states $\mathbf{k}_r$ $(k_r > 0)$ in the mixture give zero deviation and the same expectation value for **a**. Taking for the k_r pure states (the eigenstates of **k**), we see that **a** is only deviationless in the state **k**, if the latter is a pure linear combination or a mixture of linear combinations of eigenstates of **a** all with the same eigenvalue.

Because one can easily find two non-degenerate quantum operators (i.e. quantum operators with no more than one eigenstate for each eigenvalue), which have no eigenstates in common (e.g. **p** and **q**), there can be no quantum states in which all observables have zero deviation (deviationless states) [1]). Here might seem to lie the reason why the observational statements of quantum mechanics are in general of statistical character. No doubt there is some connection, but this rapid conclusion should not be taken too rashly, because it implies an interpretation of the deviation, which is not entirely justified. Let us turn to this interpretation.

In a statistical description of the 1st kind S^1 the deviation of a quantity a is defined by

$$Dev(a) = Ex((a - Ex(a))^2) = Ex(a^2) - (Ex(a))^2. \tag{2.09}$$

In an ensemble, in which this deviation is zero, a must have the same value in all samples. Then it follows that for every function $f(a)$

$$Ex(f(a)) = f(Ex(a)). \tag{2.10}$$

Whereas in general a has a proper value only in a sample and in an ensemble only an average value (expectation value), one can speak of the proper value of a in an ensemble if the deviation is zero.

In quantum mechanics it is not entirely clear what is meant by the square or another function of an observable. In order to discuss things, let us have recourse for a moment to the notion of 1.10 and let a stand for the observable represented by $\mathbf{a}(a \longleftrightarrow \mathbf{a}$; problem α_4). Then (2.09) is only identical with (2.02) for all states **k** if

$a^2 \longleftrightarrow \mathbf{a}^2$. Further we have seen that a state $\mathbf{k}$, in which (2.02) is zero, must be a (mixture of) linear combination(s) of eigenstates of $\mathbf{a}$ all with the same eigenvalue a_μ. In these states the eigenvalue of $f(\mathbf{a})$ is $f(a_\mu)$ and $Dev(\mathbf{k}; f(\mathbf{a})) = 0$. We write the operator, which represents $f(a)$ as $\mathbf{f}(a)$. If (2.10) shall be valid in a state $\mathbf{k}$, in which (2.02) is zero, we must have

$$Tr(\mathbf{k}\mathbf{f}(a)) = f(Tr(\mathbf{k}\mathbf{a})) = f(a_\mu) = Tr(\mathbf{k} f(\mathbf{a})); \quad Dev(\mathbf{k}; \mathbf{f}(a)) = 0. \ (2.11)$$

The second part is a special case of the first. The first part requires that the matrix elements of $\mathbf{f}(a)$ with respect to the eigenstates of $\mathbf{a}$ with the same eigenvalue a_μ have to be the same as those of $f(\mathbf{a})$ (i.e. equal to $f(a_\mu)$), the second part that the matrix elements of $\mathbf{f}(a)$ with respect to the eigenstates of $\mathbf{a}$ with different eigenvalues a_μ are zero like those of $f(\mathbf{a})$. This means $\mathbf{f}(a) = f(\mathbf{a})$ so that I has to be satisfied. For every a, for which I is accepted, (2.10) always holds in states in which a has zero deviation. For those a, for which I is rejected, (2.10) breaks down even in such states. In the latter case it should be kept in mind that if we speak about a_μ as the proper value of the observable $\mathbf{a}$ in such a state, this is actually more or less misleading.

Thus we could give a meaning to the deviation, as soon as we could give a meaning to problem α_4 (or the special case α_5). This meaning would only agree with the one which is usually prematurely accepted, as long as rule I would hold. From the quantummechanical point of view O_q there is no need for such a meaning. Meanwhile from the formal point of view the definiteness of the expression (2.02) remains of interest.

2.02 *The measuring device* [1]). The aim of an (ideal) measuring process is to infer (the most complete) data of the object system from the data of the observational perception. Object system and observer interact by intervention of a chain of systems, which form the measuring instrument. This chain can be cut into two parts. The first part (which may be empty) can be added to the object system, the last part to the observer. Extended object system and extended observer interact directly. The (extended) object system is regarded as a physical system. It is described by a physical treatment. The (extended) observer is unsusceptible of a physical treatment. Its part consists in an act, which must be stated without further analysis. The result of the measuring process should be independent of the place of the cut in the measuring system, provided the first part is entirely accessible to a physical treatment.

We make a simplified model of the extended object system in which all partaking systems have one degree of freedom. The original object system is denoted by 1, the successive systems of the measuring instrument before the cut by $2, 3, \ldots n$. Every pair of adjacent systems $l - 1$ and l $(l = 2, 3, \ldots n)$ is coupled during a time interval (t_{2l-4}, t_{2l-3}). The time intervals must be ordered so, that

$$t_{2k+1} \geqslant t_{2k-1}. \tag{2.12}$$

For the sake of simplicity we impose the condition that different time intervals do not overlap

$$t_k > t_{k-1}. \tag{2.13}$$

Then the couplings between the various pairs of adjacent systems can successively be treated separately.

In 1 we choose a complete system of orthonormal wave functions $\varphi'_{1\mu}(t)$. The time dependence can be described with the help of a H e r m i t i a n operator $\mathbf{H}^0_1(t)$ according to

$$-\frac{\hbar}{i} \frac{\partial}{\partial t} \varphi'_{1\mu}(t) = \mathbf{H}^0_1(t) \varphi'_{1\mu}(t). \tag{2.14}$$

1 is coupled with 2 during the time interval (t_0, t_1). This means that during this time interval the H a m i l t o n i a n $\mathbf{H}_{12}(t)$ of the combined systems 1 and 2 cannot be split up into the sum of two H a m i l t o n i a n s $\mathbf{H}_1(t)$ and $\mathbf{H}_2(t)$ of the separate systems. The system 2 is supposed to be initially in the pure quantum state $\varphi_{20}(t_0)$.

We impose two conditions on $\mathbf{H}_{12}(t)$ and $\varphi_{20}(t_0)$. The first condition is that $\mathbf{H}_{12}(t) - \mathbf{H}^0_1(t)$ must be diagonal with respect to the system of $\varphi'_{1\mu}(t)$

$$(\mathbf{H}_{12}(t) - \mathbf{H}^0_1(t)) \varphi'_{1\mu}(t) = \varphi'_{1\mu}(t) \mathbf{G}_{\mu 2}(t). \tag{2.15}$$

$\mathbf{G}_{\mu 2}$ is an operator with respect to the variables of 2 (q-number), but an ordinary number with respect to the variables of 1 (c-number).

When 1 is initially in the pure quantum state $\varphi'_{1\mu}(t_0)$, the final state of 1 and 2 together is because of the wave equation

$$\frac{\hbar}{i} \frac{\partial}{\partial t} \varphi_{12}(t) = - \mathbf{H}_{12}(t) \, \varphi_{12}(t) \tag{2.16}$$

given by

$$e^{-\frac{i}{\hbar} \int_{t_0}^{t_1} dt \, \mathbf{H}_{12}(t)} \varphi'_{1\mu}(t_0) \varphi_{20}(t_0) = \varphi'_{1\mu}(t_1) \, e^{-\frac{i}{\hbar} \int_{t_0}^{t_1} dt \mathbf{G}_{\mu 2}(t)} \varphi_{20}(t_0). \tag{2.17}$$

With arbitrary chosen functions $g_\mu(t)$ and

$$\varphi_{1\mu}(t) = \varphi'_{1\mu}(t)e^{-\frac{i}{\hbar}\int_{t_0}^{t} dt' g_\mu(t')} ;$$

$$\varphi_{2\mu}(t) = e^{-\frac{i}{\hbar}\int_{t_0}^{t} dt'(-g_\mu(t')+G_{\mu 2(t)})}\varphi_{20}(t_0) \quad (t_0 \leqslant t \leqslant t), \quad (2.18)$$

(2.17) becomes

$$\varphi_{1\mu}(t_1)\varphi_{2\mu}(t_1). \quad (2.19)$$

The second condition, which we impose on $\mathbf{H}_{12}(t)$ and $\varphi_{20}(t)$ is that the (already normalized) $\varphi_{2\mu}(t_1)$ must be orthogonal

$$\varphi^\dagger_{2\mu}(t_1)\varphi_{2\nu}(t_1) = \varphi^\dagger_{20}(t_0)\, e^{\frac{i}{\hbar}\int_{t_\nu}^{t_1} dt\, (-g_\mu(t)+G_{\mu 2(t)})}$$

$$. \, e^{-\frac{i}{\hbar}\int_{t_0}^{t_1} dt(-g_\nu(t)+G_{\nu 2(t)})}\varphi_{20}(t_0) = \delta_{\mu\nu}. \quad (2.20)$$

The system of $\varphi_{2\mu}(t_1)$ need not be complete.

For $t > t_1$, after the coupling has been dissolved, 1 and 2 have separate H a m i l t o n i a n operators $\mathbf{H}_1(t)$ and $\mathbf{H}_2(t)$. The ortho-normal functions $\varphi_{1\mu}(t_1)$ and $\varphi_{2\mu}(t_2)$ then transform into the ortho-normal functions

$$\text{and} \quad \begin{matrix} \varphi_{1\mu}(t) = e^{-\frac{i}{\hbar}\int_{t_1}^{t} dt'\, \mathbf{H}_1(t')}\varphi_{1\mu}(t_1) \\ \\ \varphi_{2\mu}(t) = e^{-\frac{i}{\hbar}\int_{t_1}^{t} dt'\, \mathbf{H}_2(t')}\varphi_{2\mu}(t_1). \end{matrix} \quad (2.21)$$

The complete wave function (2.19) transforms into

$$\varphi_{1\mu}(t)\varphi_{2\mu}(t) \quad (t \geqslant t_1). \quad (2.22)$$

The succeeding pairs of adjacent systems are coupled analogously. The complete wave function of the first m systems after the last coupling becomes, in the same way as (2.22),

$$\varphi_{1\mu}(t)\varphi_{2\mu}(t)\ldots\varphi_{m\mu}(t) \quad (t_{2m-3} \leqslant t \leqslant t_{2m-2}). \quad (2.23)$$

More general 1 can, instead of being in a pure state $\varphi_{1\mu}(t_0)$, be initially in a state with statistical operator $\mathbf{k}_1(t_0)$, which then can be expanded according to

$$\mathbf{k}_1(t_0) = \sum_{\mu,\nu} \varkappa_{1\nu\mu}(t_0)\mathbf{k}_{1\mu\nu}(t_0) \text{ with } \varkappa_{1\nu\mu}(t_0) = Tr(\mathbf{k}_{1\nu\mu}(t_0)\mathbf{k}_1(t_0)). \quad (2.24)$$

The statistical operator of the first m systems after the last interaction then becomes with (2.23)

$$\mathbf{k}_{12\ldots m}(t) = \sum_{\mu,\nu} \varkappa_{1\nu\mu}(t_0)\mathbf{k}_{1\mu\nu}(t)\mathbf{k}_{2\mu\nu}(t)\ldots\mathbf{k}_{m\mu\nu}(t) \ (t_{2m-3} \leqslant t \leqslant t_{2m-2}). \quad (2.25)$$

The interactions have affected the states of the partaking systems and established a correlation between them (entanglement).

2.03 *Infringed states*. When after the interaction the correlation between the state of an arbitrary system $l(1 \leqslant l \leqslant m)$ and the state of the other $m - 1$ of the first m systems is ignored, the latter state can irrespective of the former state according to (2.25) and (1.11) be described by the statistical operator

$$\mathbf{k}_{12\ldots(l-1)\,(l+1)\ldots m}(t) = Tr_l\,\mathbf{k}_{12\ldots m}(t)$$

$$= \sum_{\mu} \varkappa_{1\mu\mu}(t_0)\mathbf{k}_{1\mu\mu}(t)\ldots\mathbf{k}_{(l-1)\mu\mu}(t)\mathbf{k}_{(l+1)\mu\mu}(t)\ldots\mathbf{k}_{m\mu\mu}(t) \quad (2.26)$$

(Tr_l denotes the trace with respect to the variables of l). More general the state of a selected series $l_1, l_2, \ldots l_k$ $(1 \leqslant l_1 < l_2 < \ldots l_k \leqslant m)$ out of the chain of the first m systems irrespective of the state of the other systems is described by the statistical operator

$$\mathbf{k}_{l_1 l_2 \ldots l_k}(t) = \sum_{\mu} \varkappa_{1\mu\mu}(t_0)\mathbf{k}_{l_1\mu\mu}(t)\mathbf{k}_{l_2\mu\mu}(t)\ldots\mathbf{k}_{l_k\mu\mu}(t) \ (t \geqslant t_{2l_k-3}). \quad (2.27)$$

(2.27) is the statistical operator of a mixture of pure quantum states $\varphi_{l_1\mu}(t)\varphi_{l_2\mu}(t)\ldots\varphi_{l_k\mu}(t)$ with weights $\varkappa_{1\mu\mu}(t_0)$. The ignorance of the correlation with other systems has also partially destroyed the correlation between the selected systems themselves. According to the remaining correlation only individual pure quantum states $\varphi_{l\mu}(t)$ of the systems $l_1, l_2, \ldots l_k$ with the same Greek index occur together. We denote a state of a group of systems, which has come about by interaction with other, afterwards ignored, systems as an infringed state. ((2.25) is the entangled state (2.27) the infringed state).

We consider two particular cases of infringed states. First we put $m = n$ and let the selected series consist of the systems 1 and n only. (2.27) then becomes

$$\mathbf{k}_{1n}(t) = \sum_{\mu} \varkappa_{1\mu\mu}(t_0)\mathbf{k}_{1\mu\mu}(t)\mathbf{k}_{n\mu\mu}(t) \ (t > t_{2n-3}). \quad (2.28)$$

The correlation between 1 and n, which is left in this infringed state, justifies the inference that when for n the pure quantum state $\varphi_{n\mu}(t)$ is realized, the corresponding pure quantum state $\varphi_{1\mu}(t)$ (with the same μ) is realized for 1. With this inference the correlation is completely exhausted.

In the second place we put $m = n + 1$ (supposing that the inter-action between n and $n + 1$, which crosses the cut, is still accessible to a physical treatment) and select the systems $1,2,\ldots n$. Then (2.27) gives

$$\mathbf{k}_{12\ldots n}(t) = \sum_{\mu} \varkappa_{1\mu\mu}(t_0)\mathbf{k}_{1\mu\mu}(t)\mathbf{k}_{2\mu\mu}(t)\ldots\mathbf{k}_{n\mu\mu}(t) \quad (t \geqslant t_{2n-1}). \quad (2.29)$$

(2.29) determines the infringed state in which the extended object system is left after the interaction with the observer, if the state of the observer is afterwards ignored.

If in (2.29) we put $n = 1$, we get

$$\mathbf{k}_1(t) = \sum_{\mu} \varkappa_{1\mu\mu}(t_0)\mathbf{k}_{1\mu\mu}(t) \quad (t \geqslant t_1), \quad (2.30)$$

which determines the infringed state of the original object system after the interaction with the measuring instrument, irrespective of the final state of the latter (and of the observer).

2.04 *The measurement conclusion.* When the original object system and observer are connected by a measuring instrument, which consists of an unramified chain of one or more interacting systems, it follows from (2.28) that the conclusion about the original object system, which the observer can infer from his final perception, certainly cannot go further than to indicate which of the pure quantum states $\varphi_{1\mu}(t)$ is realized. According to the quantum notion of observation O_q the observer can in principle actually infer that conclusion under ideal conditions and he cannot infer more under any condition. This rule establishes the connection between the mathematical formalism and the observers perceptions. The rule does not follow from the formalism. The formalism is in harmony with the rule. The rule justifies the representation of the formalism in terms of pure quantum states.

The conclusion derived from the measurement thus consists in indicating which pure quantum state of the mixture (2.29) or (2.30) of the extended or original object system is realized after this measurement. It could indicate equally well the realized pure quantum state of an arbitrary system or group of systems of the measuring instrument. For a great number of measurements on identical object systems with identical initial operators the statistical probability of realization of a pure quantum state with index μ is according to the statistical interpretation of (2.29) or (2.30) $\varkappa_{1\mu\mu}(t_0)$ (cf. O_q). The measuring result is independent of the place of the cut in the measuring instrument [1]).

Formally we can distinguish the following stages in the measuring act. First the object system is coupled with the measuring instrument, which gives the entangled state, then the systems of the measuring chain are ignored, which gives the infringed mixture, from which finally the realized state is selected. They are represented by the scheme:

$$
\begin{array}{ccc}
 & \textit{initial state} & \mathbf{k}_1(t) = \sum_{\mu,\nu} \varkappa_{\nu\mu}\mathbf{k}_{1\mu\nu}(t) \\
\mathbf{coupling} & \downarrow & \\
 & \textit{entangled state} & \sum_{\mu,\nu} \varkappa_{\nu\mu}\mathbf{k}_{1\mu\nu}(t)\mathbf{k}_{2\mu\nu}(t)\,.... \\
\mathbf{ignoration} & \downarrow & \\
 & \textit{infringed state} & \sum_{\mu} \varkappa_{\mu\mu}\mathbf{k}_{1\mu\mu}(t) \\
\mathbf{selection} & \downarrow & \\
 & \textit{measured state} & \mathbf{k}_{1\mu\mu}(t)
\end{array}
$$

2.05 *The measuring of observables.* For every system l we can define a H e r m i t i a n operator $\mathbf{a}_l(t)$ for which the functions $\varphi_{l\mu}(t)$ form a system of orthonormal eigenfunctions with arbitrary prescribed eigenvalues $a_{l\mu}(t)$. $\mathbf{a}_l(t)$ commutes with $\mathbf{H}_l^0(t)$

$$[\mathbf{H}_i^0(t), \mathbf{a}_i(t)] = 0. \tag{2.31}$$

The condition (2.15) is then equivalent to the condition that $\mathbf{H}_{12}(t)$ must commute with $\mathbf{a}_1(t)$, or in general

$$[\mathbf{H}_{l(l+1)}(t), \mathbf{a}_l(t)] = 0. \tag{2.32}$$

In the pure quantum state $\varphi_{l\mu}(t)$ the observable $\mathbf{a}_l(t)$ has the value $a_{l\mu}(t)$. A measurement, which decides which of the states $\varphi_{l\mu}(t)$ of l is realized, also determines the value of $\mathbf{a}_l(t)$. It can be regarded as a measurement of the observable $\mathbf{a}_l(t)$. This establishes the experimental meaning of the value of an observable. Meanwhile, remembering 2.01, one should be careful in regarding $a_{l\mu}(t)$ as the proper value of $\mathbf{a}_l(t)$.

If all eigenvalues of $\mathbf{a}_l(t)$ are different

$$a_{l\mu}(t) \neq a_{l\nu}(t) \text{ for } \mu \neq \nu, \tag{2.33}$$

the value of $\mathbf{a}_l(t)$ on the other hand uniquely determines the pure quantum state of the system l. Therefore, instead of indicating which state $\varphi_{l\mu}(t)$ of l is realized, the observer can in the ideal case (2.33) equally well (and otherwise less well) record the value of

$a_l(t)$. Usually the measuring results are thus stated in terms of values of observables and not in terms of states. For this purpose it is immaterial whether these values (defined as eigenvalues) have a proper meaning in the sense of 2.01 or not.

2.06 *Correlated observables.* Similarly a correlation between the states of various systems can also be expressed as a correlation between the values of observables of these systems. As a particular case we consider the effect of ignoring the correlation with some systems (infringement) on the correlation between the remaining systems. This effect has in 2.03 been found to consist in the disappearance of the non-diagonal statistical operators $k_{l\nu\mu}(t)$ ($\mu \neq \nu$) of the latter systems. This has no influence upon the expectation values of those observables, for which the operators are diagonal with respect to the functions $\varphi_{l\mu}(t)$. That means that the correlation between such observables, for which the operators commute with the $a_l(t)$, remains unaffected. For other observables the non-diagonal elements are dropped and the correlation is more or less destroyed. For observables, for which the operator has no non-zero diagonal elements with respect to the $\varphi_{l\mu}(t)$, no elements remain and the correlation is entirely destroyed.

2.07 *The pointer reading.* When for some system in the chain, say l, the functions $\varphi_{l\mu}(t)$ read in q-representation

$$\varphi_{l\mu}(t) = \delta(q_l - q_{l\mu}), \tag{2.34}$$

so that they are eigenfunctions of q_l

$$q_l \varphi_{l\mu} = q_{l\mu} \varphi_{l\mu}, \tag{2.35}$$

we denote the measurement as a (pointer) reading. l is called the scale system. The measuring result of a reading can be expressed by the value of the coordinate of the scale system.

A simplified model, which gives such a coupling between the systems $(l-1)$ and l, that the values of the observables $a_{(l-1)}(t)$ are measured by the values of the coordinate q_l, is obtained [1]) with a H a m i l t o n i a n operator of the type

$$H_{(l-1)l}(t) = h(a_{(l-1)}(t)) + f(a_{(l-1)}(t)) p_l. \tag{2.36}$$

The condition (2.32) is satisfied. With the choice

$$g_\mu(t) = h(a_{(l-1)\mu}(t)) \tag{2.37}$$

(2.18) gives

$$\varphi_{l\mu}(t) = e^{-\frac{i}{\hbar}\int_{t_0}^{t_1} dt\, f(a_{(l-1)\mu}(t))\mathbf{p}_l}\, \varphi_{l0}(t_0).\tag{2.38}$$

We suppose that the wave function of the initial state of l reads in q_l-representation

$$\varphi_{l0}(q_l;\, t_0) = \delta(q_l - q_{l0}),\tag{2.39}$$

so that $\mathbf{q}_l$ has the initial value q_{l0}

$$\mathbf{q}_l\varphi_{l0}(t_0) = q_{l0}\varphi_{l0}(t_0).\tag{2.40}$$

(2.38) then gives

$$\varphi_{l\mu}(t_1) = \delta(q_l - q_{l0} - F(a_{(l-1)\mu}));\; F(a_{(l-1)\mu}) = \int_{t_0}^{t_1} dt\, f(a_{(l-1)\mu}(t)).\tag{2.41}$$

If we put

$$q_{l\mu} = q_{l_0} - F(a_{(l-1)\mu}),\tag{2.42}$$

(2.41) becomes

$$\varphi_{l\mu}(t_1) = \delta(q_l - q_{l\mu}).\tag{2.43}$$

These wave functions are eigenfunctions of $\mathbf{q}_l$ with eigenvalues $q_{l\mu}$

$$\mathbf{q}_l\varphi_{l\mu}(t_1) = q_{l\mu}\varphi_{l\mu}(t_1).\tag{2.44}$$

The orthogonality condition (2.20) requires

$$q_{l\mu} \neq q_{l\nu} \text{ for } \mu \neq \nu,\tag{2.45}$$

which is at the same time equivalent to the condition (2.33). (2.45) is satisfied if

$$F(a_{(l-1)\mu}) \neq F(a_{(l-1)\nu}) \text{ for } \mu \neq \nu.\tag{2.46}$$

The spectrum of the values $q_{l\mu}$ (2.42) need not necessarily cover the whole domain of values of $\mathbf{q}_l$ from $-\infty$ until $+\infty$.

The momentum operator $\mathbf{p}_l$ reads in q_l-representation

$$\mathbf{p}_l = \frac{\hbar}{i}\frac{\partial}{\partial q_l}.\tag{2.47}$$

The matrix elements with respect to the functions (2.43) are

$$Tr(\mathbf{p}_l\mathbf{k}_{l\nu\mu}) = \frac{\hbar}{i}\frac{\partial}{\partial q_{l\nu}}\delta(q_{l\nu} - q_{l\mu}).\tag{2.48}$$

The diagonal elements ($\mu = \nu$) are zero. Therefore the correlation of the momentum $\mathbf{p}_l$ of the scale system with observables of other systems is entirely destroyed by the measurement of the canonical conjugate coordinate $\mathbf{q}_l$.

2.08 *Observational connections.* For a relation between observational data we need at least two measurements. We consider two succeeding measurements during the time intervals (t_0,t_1) and (t_0',t_1') with

$$t_0' > t_1 \qquad (2.49)$$

performed on the same system 1. The first measurement measures the states $\varphi_{1\mu}(t)$ or a corresponding observable $\mathbf{a}_1(t)$, the second one measures the states $\varphi_{1\mu}'(t)$ or a corresponding observable $\mathbf{a}_1'(t)$.

If the first measuring result indicates the final pure quantum state $\varphi_{1\mu}(t)$ $(t_1 \leqslant t \leqslant t_0')$, the statistical operator at the beginning t_0' of the second measurement is $\mathbf{k}_{1\mu\mu}(t_0')$, which is expanded according to

$$\mathbf{k}_{1\mu\mu}(t_0') = \sum_{\mu',\nu'} \varkappa_{1\mu\mu,\nu'\mu'}'(t_0') \, \mathbf{k}_{1\mu'\nu'}'(t_0')$$

with

$$\varkappa_{1\mu\mu,\nu'\mu'}'(t_0') = Tr(\mathbf{k}_{1\nu'\mu'}'(t_0')\mathbf{k}_{1\mu\mu}(t_0')).$$

$$(2.50)$$

The statistical probability, that, after the first measuring result has indicated the pure quantum state $\varphi_{1\mu}(t)$ $(t_1 \leqslant t \leqslant t_0')$, the second measuring result will indicate the pure quantum state $\varphi_{1\nu}'(t)$ $(t \geqslant t_1')$ is

$$\varkappa_{1\mu\mu,\nu'\nu}'(t_0') = Tr(\mathbf{k}_{1\nu'\nu}'(t_0') \, \mathbf{k}_{1\mu\mu}(t_0')) = \mid \varphi_{1\nu}'^{\dagger}(t_0') \, \varphi_{1\mu}(t_0') \mid^2. \quad (2.51)$$

This conditional probability is actually the most elementary expression contained in the formalism, which denotes an observable connection and which has a directly observable statistical meaning.

When the functions $\varphi_{1\mu}'(t)$ coincide with the $\varphi(t_{1\mu})$, i.e. when $\mathbf{a}_1'(t)$ and $\mathbf{a}_1(t)$ commute, (2.51) becomes

$$\varkappa_{1\mu\mu,\nu'\nu}'(t_0') = \delta_{\nu'\mu} \qquad (2.52)$$

and the second measuring result can be predicted with certainty from the first. In this case we have essentially the repetition of a measurement. (2.52) expresses the reproducibility of the measuring result.

2.09 *Intermingled states.* The entangled state of two object systems 1 and 2 after a coupling of the type described above is of the kind

$$\mathbf{k}_{12} = \sum_{\mu,\nu} \varkappa_{\nu\mu} \, \mathbf{k}_{1\mu\nu} \, \mathbf{k}_{2\mu\nu}. \qquad (2.53)$$

The probability of finding system 1 in a state $\mathbf{k}_1$ and 2 in a state $\mathbf{k}_2$ is

$$Tr(\mathbf{k}_{12}\mathbf{k}_1\mathbf{k}_2) = \sum_{\mu,\nu} \varkappa_{\nu\mu} \, Tr(\mathbf{k}_{1\mu\nu}\mathbf{k}_1) \, Tr(\mathbf{k}_{2\mu\nu}\mathbf{k}_2). \qquad (2.54)$$

When $\mathbf{k}_1$ and $\mathbf{k}_2$ coincide with the projection operators $\mathbf{k}_{1\mu\mu}$ and $\mathbf{k}_{2\nu\nu}$,

(2.54) becomes equal to $\varkappa_{\mu\mu}\delta_{\mu\nu}$. This might (wrongly) suggest that (after the coupling and before the measurement) the state of 1 and 2 is the mixture

$$\mathbf{k}'_{12} = \sum_{\mu} \varkappa_{\mu\mu} \mathbf{k}_{1\mu\mu} \, \mathbf{k}_{2\mu\mu} \tag{2.55}$$

instead of the state (2.53). In this way the correlation between 1 and 2 would be partially destroyed by the omission of the non-diagonal terms. In the mixture (2.55) the expectation value of the states $\mathbf{k}_1$ and $\mathbf{k}_2$ would be

$$Tr(\mathbf{k}'_{12}\mathbf{k}_1\mathbf{k}_2) = \sum_{\mu} \varkappa_{\mu\mu} \, Tr(\mathbf{k}_{1\mu\mu} \, \mathbf{k}_1) \, Tr(\mathbf{k}_{2\mu\mu} \, \mathbf{k}_2) \tag{2.56}$$

instead of (2.54). It has been emphasized by F u r r y [3]) (in a somewhat different form and particularly against our common opponents, cf. 2.11) that only if neither $\mathbf{k}_1$ nor $\mathbf{k}_2$ coincides with any of the $\mathbf{k}_{1\mu\mu}$ or $\mathbf{k}_{2\nu\nu}$ respectively, (2.56) can be different from (2.54). Because the latter case hardly occurs in the relevant applications, one is apt to make the mistake of replacing (2.53) by (2.55) (and to draw unjustified conclusions whenever this case does occur).

If 1 and 2 had been coupled with one or more further systems 3,.... according to

$$\mathbf{k}_{123....} = \sum_{\mu,\nu} \varkappa_{\nu\mu} \, \mathbf{k}_{1\mu\nu} \, \mathbf{k}_{2\mu\nu} \, \mathbf{k}_{3\mu\nu} \cdots \tag{2.57}$$

and these further systems had been ignored afterwards, the infringed state of 1 and 2 would correctly be given by (2.55) indeed. This infringed state is quite distinct from the entangled state (2.53).

2.10 *Multilateral correlation.* In (2.53) the transition operators $\mathbf{k}_{1\mu\nu}$ and $\mathbf{k}_{2\mu\nu}$ belong to two systems of orthonormal wave functions $\varphi_{1\mu}$ and $\varphi_{2\mu}$, which span the (generalized) H i l b e r t subspaces R_1 and R_2. An interesting case [4]) is that for which $\mathbf{k}_{12}$ can similar to (2.53) also be expanded with respect to the transition operators $\mathbf{l}_{1\rho\sigma}$ and $\mathbf{l}_{2\rho\sigma}$ belonging to any two systems of wave functions $\psi_{1\rho}$ and $\psi_{2\rho}$ in R_1 and R_2, when one system is chosen arbitrarily variable but orthonormal and complete, the other system suitably to the first

$$\sum_{\mu,\nu} \varkappa_{\nu\mu} \, \mathbf{k}_{1\mu\nu} \, \mathbf{k}_{2\mu\nu} = \sum_{\rho,\sigma} \lambda_{\sigma\rho} \, \mathbf{l}_{1\rho\sigma} \, \mathbf{l}_{2\rho\sigma}. \tag{2.58}$$

A necessary and sufficient condition [4]) for the occurrence of this case is that the $\varkappa_{\nu\mu}$ are of the form

$$\varkappa_{\nu\mu} = \varkappa_{\nu}^* \, \varkappa_{\mu}; \, |\varkappa_{\mu}| = \varkappa. \tag{2.59}$$

The factorization of $\varkappa_{\nu\mu}$ means that $\mathbf{k}_{12}$ is a pure quantum state of the combined systems 1 and 2 with wave function

$$\varphi_{12} = \sum_{\mu} \varkappa_{\mu}\, \varphi_{1\mu}\, \varphi_{2\mu}. \tag{2.60}$$

The unimodular coefficients $\varkappa_{\mu}/\varkappa$ could even be included in $\varphi_{1\mu}$ or $\varphi_{2\mu}$.

The special case under discussion can easily be generalized to the following case. The functions $\varphi_{1\mu}$ and $\varphi_{2\mu}$ are taken together in groups $\varphi_{1\mu_1}, \varphi_{1\mu_2}, \ldots$ and $\varphi_{2\mu_1}, \varphi_{2\mu_2}, \ldots$, which span the (generalized) H i l b e r t subspaces $R_{11}, R_{12}, \ldots$ and $R_{21}, R_{22}, \ldots$ respectively $(R_1 = R_{11} + R_{12} + \ldots$ and $R_2 = R_{21} + R_{22} + \ldots)$. In these subspaces we take any two sets of systems $\psi_{1\rho_1}, \psi_{1\rho_2}, \ldots$ and $\psi_{2\rho_1}, \psi_{2\rho_2}, \ldots$, of which one set is chosen arbitrarily variable but orthonormal and complete, the other suitably to the first. It is easily seen that the last part of condition (2.59) then has to be replaced by $|\varkappa_{\mu_p}| = \varkappa_p$. In 1-dimensional subspaces R_{1p} and R_{2p} all 1-representations are essentially the same.

An equivalent formulation of the generalized case is obtained by taking instead of any two systems of wave functions $\psi_{1\rho}$ and $\psi_{2\rho}$, as in the special case, two definite systems of which one is chosen arbitrarily fixed but orthonormal and complete, the other suitably to the first. $R_{11}, R_{12}, \ldots$ or $R_{21}, R_{22}, \ldots$ are then determined by the sharpest division of R_1 or R_2 into subspaces, which span linearly independent groups of $\varphi_{1\mu}$ and $\psi_{1\rho}$ or $\varphi_{2\mu}$ and $\psi_{2\rho}$ at the same time.

We restrict ourselves to the special case. First we show the necessity of (2.59). With (1.13) it follows from (2.58) that

$$\varkappa_{\nu\mu}\, Tr_1(\mathbf{k}_{1\mu\nu}\, \mathbf{l}_{1\sigma\rho}) = \lambda_{\sigma\rho} Tr_2(\mathbf{k}_{2\nu\mu}\, \mathbf{l}_{2\rho\sigma}),$$
$$\varkappa_{\nu\mu} Tr_2(\mathbf{k}_{2\mu\nu}\, \mathbf{l}_{2\rho\sigma}) = \lambda_{\sigma\rho} Tr_1(\mathbf{k}_{1\nu\mu}\, \mathbf{l}_{1\rho\sigma}). \tag{2.61}$$

It follows directly that

$$\varkappa_{\mu\nu}\varkappa_{\nu\mu}\, Tr_l(\mathbf{k}_{l\mu\nu}\, \mathbf{l}_{l\sigma\rho}) = \lambda_{\sigma\rho}\, \lambda_{\rho\sigma}\, Tr_l\, (\mathbf{k}_{l\mu\nu}\, \mathbf{l}_{l\sigma\rho})\ (l = 1,2), \tag{2.62}$$

so that (with $\varkappa_{\mu\nu} = \varkappa_{\nu\mu}^*$, $\lambda_{\rho\sigma} = \lambda_{\sigma\rho}^*$)

$$|\varkappa_{\mu\nu}|^2 = |\lambda_{\rho\sigma}|^2 \text{ or } Tr_l\, (\mathbf{k}_{l\mu\nu}\, \mathbf{l}_{l\sigma\rho}) = 0\ (l = 1 \text{ and } 2). \tag{2.63}$$

Because one of the systems $\mathbf{l}_{1\rho\sigma}$ or $\mathbf{l}_{2\rho\sigma}$ is arbitrarily variable and complete in R_1 or R_2 the latter alternative is excluded and we must have

$$|\varkappa_{\mu\nu}| = |\lambda_{\rho\sigma}| = \varkappa^2 = \lambda^2\ (\varkappa = \lambda > 0). \tag{2.64}$$

With (1.13) it further follows from (2.58) that

$$\sum_{\mu,\nu} \varkappa_{\nu\mu} T\gamma_1(\mathbf{k}_{1\mu\nu}\,\mathbf{l}_{1\sigma\rho})\,\mathbf{k}_{2\mu\nu} = \lambda_{\sigma\rho}\,\mathbf{l}_{2\rho\sigma},$$

$$\sum_{\mu,\nu} \varkappa_{\nu\mu} T\gamma_2(\mathbf{k}_{2\mu\nu}\,\mathbf{l}_{2\sigma\rho})\,\mathbf{k}_{1\mu\nu} = \lambda_{\sigma\rho}\,\mathbf{l}_{1\rho\sigma}.$$

(2.65)

These relations connect the arbitrarily and the suitably chosen systems and establish the orthonormality and completeness of the latter. With (1.08) we derive from (2.65)

$$\mathbf{l}_{2\rho\sigma}\,\mathbf{l}_{2\sigma'\rho'} = \frac{1}{\lambda_{\sigma\rho}\,\lambda_{\rho'\sigma'}} \sum_{\mu,\nu,\mu'} \varkappa_{\nu\mu}\,\varkappa_{\mu'\nu}\, T\gamma_1(\mathbf{k}_{1\mu\nu}\,\mathbf{l}_{1\sigma\rho})T\gamma_1(\mathbf{k}_{1\nu\mu'}\,\mathbf{l}_{1\rho'\sigma'})\,\mathbf{k}_{2\mu\mu'} \quad (2.66)$$

and

$$\mathbf{l}_{2\rho\rho'}\,\delta_{\sigma\sigma'} = \frac{1}{\lambda_{\rho\rho'}} \sum_{\mu,\nu,\mu'} \varkappa_{\mu'\mu}\, T\gamma_1(\mathbf{k}_{1\mu\nu}\,\mathbf{l}_{1\sigma\rho})\, T\gamma_1(\mathbf{k}_{1\nu\mu'}\,\mathbf{l}_{1\rho'\sigma'})\,\mathbf{k}_{2\mu\mu'} \quad (2.67)$$

and similarly for interchanged indices 1 and 2. (2.66) and (2.67) must be identical according to (1.08). Because one of the systems $\mathbf{l}_{1\rho\sigma}$ or $\mathbf{l}_{2\rho\sigma}$ is arbitrarily variable and complete in R_1 or R_2, we must have (remembering (2.64))

$$\varkappa_{\mu'\nu}\,\varkappa_{\nu\mu} = \varkappa^2\,\varkappa_{\mu'\mu};\; \lambda_{\rho'\sigma}\,\lambda_{\sigma\rho} = \lambda^2\,\lambda_{\rho'\rho}\;(\varkappa = \lambda > 0). \quad (2.68)$$

Then $\varkappa_{\nu\mu}$ and $\lambda_{\rho\sigma}$ must have the form

$$\varkappa_{\nu\mu} = \varkappa_\nu^*\,\varkappa_\mu,\; |\,\varkappa_\mu\,| = \varkappa;\; \lambda_{\sigma\rho} = \lambda_\sigma^*\,\lambda_\rho,\; |\,\lambda_\rho\,| = \lambda. \quad (2.69)$$

This shows the necessity of (2.59).

The sufficiency can be shown in the following way. Choose, say in R_1, a complete system of orthonormal wave functions $\psi_{1\rho}$ and choose for each ρ a constant λ_ρ with $|\,\lambda_\rho\,| = \lambda = \varkappa$. Then take the functions

$$\psi_{2\rho} = \frac{1}{\lambda_\rho} \sum_\mu \varkappa_\mu\,(\psi_{1\rho}^\dagger\,\varphi_{1\mu})\,\varphi_{2\mu}, \quad (2.70)$$

which are orthonormal and complete in R_2. From (2.70) it follows that

$$\psi_{1\rho} = \lambda_\rho \sum_\mu \frac{1}{\varkappa_\mu}\,(\psi_{2\rho}^\dagger\,\varphi_{2\mu})\,\varphi_{1\mu}. \quad (2.71)$$

The indices 1 and 2 could equally well have been interchanged. For the transition operators we get

$$\mathbf{l}_{2\rho\sigma} = \frac{1}{\lambda_{\sigma\rho}} \sum_{\mu,\nu} \varkappa_{\nu\mu}\, T\gamma_1(\mathbf{k}_{1\mu\nu}\,\mathbf{l}_{1\sigma\rho})\,\mathbf{k}_{2\mu\nu},$$

$$\mathbf{l}_{1\rho\sigma} = \lambda_{\sigma\rho} \sum_{\mu,\nu} \frac{1}{\varkappa_{\nu\mu}}\, T\gamma_2(\mathbf{k}_{2\mu\nu}\,\mathbf{l}_{2\sigma\rho})\,\mathbf{k}_{1\mu\nu}$$

(2.72)

and

$$\varkappa_{\nu\mu} Tr_1(\mathbf{k}_{1\mu\nu} \mathbf{l}_{1\sigma\rho}) = \lambda_{\sigma\rho} Tr_2(\mathbf{k}_{2\nu\mu} \mathbf{l}_{2\rho\sigma}). \qquad (2.73)$$

Therefore

$$\sum_{\mu,\nu} \varkappa_{\nu\mu} \mathbf{k}_{1\mu\nu} \mathbf{k}_{2\mu\nu} = \sum_{\mu,\nu;\,\rho,\sigma} \varkappa_{\nu\mu} Tr_1 (\mathbf{k}_{1\mu\nu} \mathbf{l}_{1\sigma\rho}) \mathbf{l}_{1\rho\sigma} \mathbf{k}_{2\mu\nu}$$

$$= \sum_{\mu,\nu;\,\rho,\sigma} \lambda_{\sigma\rho} Tr_2(\mathbf{k}_{2\nu\mu} \mathbf{l}_{2\sigma\rho}) \mathbf{l}_{1\rho\sigma} \mathbf{k}_{2\mu\nu} = \sum_{\rho,\sigma} \lambda_{\sigma\rho} \mathbf{l}_{1\rho\sigma} \mathbf{l}_{2\rho\sigma}. \qquad (2.74)$$

This shows the sufficiency of (2.59).

It is of importance for the discussion of the measuring process, that (contrary to the expectation of R u a r k [5]) multilateral correlation between more than two systems is impossible. We first show this impossibility for the case of 3 systems.

Suppose we would have the expansions

$$\mathbf{k}_{123} = \sum_{\mu,\nu} \varkappa_{\nu\mu} \mathbf{k}_{1\mu\nu} \mathbf{k}_{2\mu\nu} \mathbf{k}_{3\mu\nu} = \sum_{\rho,\sigma} \lambda_{\sigma\rho} \mathbf{l}_{1\rho\sigma} \mathbf{l}_{2\rho\sigma} \mathbf{l}_{3\rho\sigma}. \qquad (2.75)$$

With (1.13) it follows from (2.75) that

$$\varkappa_{\nu\mu} Tr_1(\mathbf{k}_{1\mu\nu} \mathbf{l}_{1\sigma\rho}) Tr_2(\mathbf{k}_{2\mu\nu} \mathbf{l}_{2\sigma\rho}) = \lambda_{\sigma\rho} Tr_3 (\mathbf{k}_{3\nu\mu} \mathbf{l}_{3\rho\sigma}) \; (cycl.),$$
$$\varkappa_{\nu\mu} Tr_3(\mathbf{k}_{3\mu\nu} \mathbf{l}_{3\sigma\rho}) = \lambda_{\sigma\rho} Tr_1(\mathbf{k}_{1\nu\mu} \mathbf{l}_{1\rho\sigma}) Tr_2(\mathbf{k}_{2\nu\mu} \mathbf{l}_{2\rho\sigma}) \; (cycl.). \qquad (2.76)$$

In the same way as before it follows that

$$| \varkappa_{\mu\nu} |^2 = | \lambda_{\rho\sigma} |^2 \; \text{ or } \; Tr_l(\mathbf{k}_{l\mu\nu} \mathbf{l}_{l\sigma\rho}) = 0 \; (l = 1, 2 \text{ and } 3). \quad (2.77)$$

Because one of the systems $\mathbf{l}_{l\rho\sigma}$ must be arbitrarily variable and complete in R_l, we must have

$$| \varkappa_{\mu\nu} | = | \lambda_{\rho\sigma} | = \varkappa^2 = \lambda^2 \; (\varkappa = \lambda > 0). \qquad (2.78)$$

It further follows from (2.76) that

$$Tr_3(\mathbf{k}_{3\mu\nu} \mathbf{l}_{3\sigma\rho}) Tr_3(\mathbf{k}_{3\nu\mu} \mathbf{l}_{3\rho\sigma}) = 1$$
or
$$Tr_1(\mathbf{k}_{1\mu\nu} \mathbf{l}_{1\sigma\rho}) Tr_2(\mathbf{k}_{2\nu\mu} \mathbf{l}_{2\rho\sigma}) = 0 \; (cycl.). \qquad (2.79)$$

Then we must have

$$Tr_1(\mathbf{k}_{1\mu\nu} \mathbf{l}_{1\sigma\rho}) = Tr_2(\mathbf{k}_{2\mu\nu} \mathbf{l}_{2\sigma\rho}) = Tr_3(\mathbf{k}_{3\mu\nu} \mathbf{l}_{3\sigma\rho}) = 1 \text{ or } 0. \quad (2.80)$$

This would mean that the systems of $\mathbf{l}_{1\rho\sigma}$, $\mathbf{l}_{2\rho\sigma}$ and $\mathbf{l}_{3\rho\sigma}$ should (but for a simultaneous change of enumeration of the Greek indices of the three corresponding operators and but for unimodular constants) be identical with those of $\mathbf{k}_{1\mu\nu}$, $\mathbf{k}_{2\mu\nu}$ and $\mathbf{k}_{3\mu\nu}$. This is against the assumption. Multilateral correlation between the states of 1, 2 and 3 is therefore impossible.

For more systems 1, 2, 3,.... the impossibility of multilateral correlation can easier be shown in the following way. Suppose we would have the expansions

$$\mathbf{k}_{123....} = \sum_{\mu,\nu} \varkappa_{\nu\mu} \mathbf{k}_{1\mu\nu} \mathbf{k}_{2\mu\nu} \mathbf{k}_{3\mu\nu} .... = \sum_{\rho,\sigma} \lambda_{\sigma\rho} \mathbf{l}_{1\rho\sigma} \mathbf{l}_{2\rho\sigma} \mathbf{l}_{3\rho\sigma} .... \quad (2.81)$$

Then

$$T\gamma_{34....} \mathbf{k}_{123....} = \sum_{\mu} \varkappa_{\mu\mu} \mathbf{k}_{1\mu\mu} \mathbf{k}_{2\mu\mu} = \sum_{\rho} \lambda_{\rho\rho} \mathbf{l}_{1\rho\rho} \mathbf{l}_{2\rho\rho}. \quad (2.82)$$

Similar to (2.61) and (2.62) we get

$$\begin{aligned} \varkappa_{\mu\mu} T\gamma_1(\mathbf{k}_{1\mu\mu} \mathbf{l}_{1\rho\rho}) &= \lambda_{\rho\rho} T\gamma_2(\mathbf{k}_{2\mu\mu} \mathbf{l}_{2\rho\rho}), \\ \varkappa_{\mu\mu} T\gamma_2(\mathbf{k}_{2\mu\mu} \mathbf{l}_{2\rho\rho}) &= \lambda_{\rho\rho} T\gamma_1(\mathbf{k}_{1\mu\mu} \mathbf{l}_{1\rho\rho}) \end{aligned} \quad (2.83)$$

and

$$\varkappa_{\mu\mu}^2 T\gamma_l(\mathbf{k}_{l\mu\mu} \mathbf{l}_{l\rho\rho}) = \lambda_{\rho\rho}^2 T\gamma_l(\mathbf{k}_{l\mu\mu} \mathbf{l}_{l\rho\rho}) \ (l = 1,2), \quad (2.84)$$

so that

$$\varkappa_{\mu\mu} = \pm \lambda_{\rho\rho} \ \text{ or } \ T\gamma_l(\mathbf{k}_{l\mu\mu} \mathbf{l}_{l\rho\rho}) = 0 \ (l = 1 \text{ and } 2). \quad (2.85)$$

Because one of the systems $\mathbf{l}_{l\rho\rho}$ is arbitrarily variable the latter alternative is excluded and because the traces in (2.83) are non-negative we must have

$$\varkappa_{\mu\mu} = \lambda_{\rho\rho}. \quad (2.86)$$

Further we have similar to (2.65)

$$\begin{aligned} \sum_{\mu} T\gamma_1(\mathbf{k}_{1\mu\mu} \mathbf{l}_{1\rho\rho}) \mathbf{k}_{2\mu\mu} &= \mathbf{l}_{2\rho\rho}, \\ \sum_{\mu} T\gamma_2(\mathbf{k}_{2\mu\mu} \mathbf{l}_{2\rho\rho}) \mathbf{k}_{1\mu\mu} &= \mathbf{l}_{1\rho\rho}, \end{aligned} \quad (2.87)$$

from which we derive

$$\mathbf{l}_{1\rho\rho} \mathbf{l}_{1\sigma\sigma} = \sum_{\mu} T\gamma_1(\mathbf{k}_{1\mu\mu} \mathbf{l}_{1\rho\rho}) T\gamma_1(\mathbf{k}_{1\mu\mu} \mathbf{l}_{1\sigma\sigma}) \mathbf{k}_{2\mu\mu} \quad (2.88)$$

and

$$\mathbf{l}_{1\rho\rho} \delta_{\rho\sigma} = \sum_{\mu} T\gamma_1(\mathbf{k}_{1\mu\mu} \mathbf{l}_{1\rho\rho}) \delta_{\rho\sigma} \mathbf{k}_{2\mu\mu} \quad (2.89)$$

and similarly for interchanged indices 1 and 2. Because (2.88) and (2.89) have to be identical according to (1.08) we must have

$$T\gamma_1(\mathbf{k}_{1\mu\mu} \mathbf{l}_{1\rho\rho}) T\gamma_1(\mathbf{k}_{1\mu\mu} \mathbf{l}_{1\sigma\sigma}) = T\gamma_1(\mathbf{k}_{1\mu\mu} \mathbf{l}_{1\rho\rho}) \delta_{\rho\sigma}. \quad (2.90)$$

This would require

$$T\gamma_1(\mathbf{k}_{1\mu\mu} \mathbf{l}_{1\rho\rho}) = \delta_{\rho\sigma} \quad (2.91)$$

for every μ, ρ and σ, which is impossible. Multilateral correlation cannot extend over more than two systems.

The proofs given for the special case of multilaterial correlation in the entire spaces $R_1, R_2, \ldots$ can easily be generalized to the general case of multilateral correlation in the subspaces $R_{11}, R_{21}, \ldots$; $R_{12}, R_{22}, \ldots; \ldots$ only.

Now we see that also in the measuring process multilateral correlation (in the special or in the generalized sense) cannot be transmitted through the chain of systems of the measuring instrument. The correlation (2.28) is uniquely determined. This excludes the possibility of surpassing in the measurement conclusion the maximum inference discussed in 2.04 by the application of multilateral correlation.

2.11 *Einstein's paradox.* We return to the two object systems 1 and 2 in the multilateral correlated state (2.58).

If the state of one of the systems, say 2, is entirely ignored, the infringed state of 1 becomes

$$x^2 \sum_{\mu} \mathbf{k}_{1\mu\mu} = \lambda^2 \sum_{\rho} \mathbf{l}_{1\rho\rho}. \qquad (2.92)$$

The sums (which are identical) denote the projection operator of the (generalized) Hilbert subspace R_1. In the mixture (2.92) all states in R_1 have the same probability $x^2 = \lambda^2$. If R_1 coincides with the entire (generalized) Hilbert space of wave functions of 1, the infringed state (2.92) becomes entirely undetermined.

If in dealing with the entangled state (2.58) one would make the mistake pointed out by Furry (cf. 2.09), one would get

$$x^2 \sum_{\mu} \mathbf{k}_{1\mu\mu} \mathbf{k}_{2\mu\mu} = \lambda^2 \sum_{\rho} \mathbf{l}_{1\rho\rho} \mathbf{l}_{2\rho\rho}. \qquad (2.93)$$

In dealing with (2.82) we have seen that (2.93) cannot hold. (2.85) does not express a correlation between pure quantum states of 1 and pure quantum states of 2 (in the way a member of (2.93) would do).

If, however, (after the interaction between 1 and 2, which establishes the state (2.58)) one of the systems, say 2, interacts with a measuring instrument, which measures the states $\mathbf{l}_{2\rho\rho}$, the infringed state of 1 and 2 together after the latter interaction is

$$\lambda^2 \sum_{\rho} \mathbf{l}_{1\rho\rho} \mathbf{l}_{2\rho\rho}. \qquad (2.94)$$

This mixture is different for different types of measurements, i.e. for different systems $\mathbf{l}_{2\rho\rho}$. (2.94) does express a correlation between pure states of 1 and pure states of 2. This correlation is of unilateral type. When the measuring result selects for 2 the state $l_{2\rho\rho}$, the state of 1 is $l_{1\rho\rho}$.

After the interaction between 1 and 2 has taken place, an observable $\mathbf{b}_1$ of 1 with eigenstates $l_{1\rho\rho}$ can be measured in two different ways: either by a direct measurement on 1, or by measuring an observable $\mathbf{b}_2$ of 2 with eigenstates $l_{2\rho\rho}$ (corresponding to $l_{1\rho\rho}$) by a direct measurement on 2 (then 2 can be conceived as a part of the measuring chain). At a first glance it might seem surprising and perhaps even paradoxical that it is still possible to decide which observable of 1 will be measured by a measurement on 2 after all interaction with 1 has been abolished [6]) and that it is possible to measure independently two incommensurable observables $\mathbf{a}_1$ and $\mathbf{b}_1([\mathbf{a}_1,\mathbf{b}_1] \neq 0)$ by applying the two measuring methods side by side [7]) [4]). (Of course one should care for not making the mistake of (2.93), which would naturally lead to paradoxical results).

When the eigenstates of $\mathbf{a}_1$ are $k_{1\mu\mu}$ and those of $\mathbf{b}_1$ are $l_{1\rho\rho}$, a measurement of $\mathbf{a}_1$ selects a state out of the left member, a measurement of $\mathbf{b}_1$ selects a state out of the right member of the expression (2.92) for the infringed state of 1. The probability that one measurement selects the state $k_{1\mu\mu}$, if the other selects the state $l_{1\rho\rho}$ (or opposite) is according to (2.51)

$$Tr(k_{1\mu\mu}l_{1\rho\rho}), \tag{2.95a}$$

no matter whether $\mathbf{a}_1$ and $\mathbf{b}_1$ are both (successively) measured directly on 1 or (no matter whether successively or simultaneously) one of them on 1 and the other one on 2. When both are directly measured on 1, the state in which 1 is left after the succeeding measurements is $k_{1\mu\mu}$ if the final measurement was that of $\mathbf{a}_1$, it is $l_{1\rho\rho}$ if the final measurement was that of $\mathbf{b}_1$. A paradoxical situation seems to arise if one asks in which state 1 is left after $\mathbf{a}_1$ has been measured on 1 and $\mathbf{b}_1$ on 2 (or opposite). We have to remember (cf. 2.08) that all observational statements bear on connections between measurements. The state in which 1 is left has only an observational meaning with regard to a succeeding measurement of an observable of 1, say c_1 with eigenstates $m_{1\tau\tau}$. When the measurement of $\mathbf{a}_1$ has selected the state $k_{1\mu\mu}$, the probability that the measurement of c_1 will select the state $m_{1\tau\tau}$ is

$$Tr(k_{1\mu\mu}m_{1\tau\tau}). \tag{2.95b}$$

When the measurement of b_1 has selected the state $l_{1\rho\rho}$, the probability that the measurement of c_1 will select the state $m_{1\tau\tau}$ is

$$Tr(l_{1\rho\rho}m_{1\tau\tau}). \tag{2.95c}$$

Thus we get two different probabilities for the same event. This is not unfamiliar in statistics, because the probabilities are (always) conditional. They have only a meaning for a great number of combined measurements of a_1, b_1 and c_1. The probability of finding a state $k_{1\mu\mu}$ is $\varkappa^2$, the probability of finding a state $l_{1\rho\rho}$ is λ^2, the probability of finding a state $m_{1\tau\tau}$ is then according to (2.95b) or (2.95c)

$$\varkappa^2 \sum_{\mu} Tr(k_{1\mu\mu}\,m_{1\tau\tau}) \text{ or } \lambda^2 \sum_{\rho} Tr(l_{1\rho\rho}\,m_{1\tau\tau}). \tag{2.96}$$

Only these sums have to be identical and they are so according to (2.92). The correlations between the measuring results for a_1, b_1 and c_1 are described by (2.95).

Let us consider once more the measurement of a_1 and of b_1, one of them directly on 1 and the other directly on 2. The latter measurement can also be conceived as a direct measurement on 1 (the system 2 is then regarded as a part of the measuring chain), which preceedes the first mentioned measurement. The only pecularity of the present case is that after the coupling between the object system 1 and the first system 2 of the measuring chain of the earliest measurement has been abolished (and even after the succeeding measurement has been performed) one can thanks to the multilateral correlation between 1 and 2 still decide which observable will be measured by this earliest measurement. But when we pay due regard to the correlations between the various measuring results, this leads to no paradox.

An illustrative example, which has been discussed by E i n-
s t e i n a.o. [7] [4]) and by B o h r a.o. [8]) [3]) [5]), is that of two particles (each with one linear degree of freedom) in an entangled state for which the wave function reads in q-representation

$$\varphi_{12} = \frac{1}{\sqrt{h}}\,\delta(q_1 - q_2 + Q)\,e^{\frac{i}{\hbar}\frac{q_1+q_2}{2}P}. \tag{2.97}$$

This state can be realized by two particles 1 and 2 directly after passing through two parallel slits at a distance Q in a diaphragm. (2.97) describes the motion in the direction perpendicular to the slits, parallel to the diaphragm. The total momentum P can be determined from the total momentum directly before the passage through the diaphragm and the change of momentum of the diaphragm. The slits can be taken so far apart, that exchange effects can be neglected.

(2.97) is of the form (2.60) with (2.59), as can be seen by expanding (2.97) with respect to e.g. coordinate or momentum eigenfunctions of 1 and 2

$$\varphi_{12} = \frac{1}{\sqrt{h}} \int d\xi \, e^{\frac{i}{\hbar} \xi P} \, \delta\left(q_1 - \xi + \frac{Q}{2}\right) \delta\left(q_2 - \xi - \frac{Q}{2}\right)$$

$$= \frac{1}{h\sqrt{h}} \int d\eta \, e^{\frac{i}{\hbar} \eta Q} \, e^{\frac{i}{\hbar} q_1 \left(\eta + \frac{P}{2}\right)} e^{\frac{i}{\hbar} q_2 \left(-\eta + \frac{P}{2}\right)}. \qquad (2.98)$$

R_1 coincides with the entire (generalized) **Hilbert** space of wave functions of 1. The infringed state of 1 is entirely undertermined. After a measuring result $q_2 = q_{2\mu}$ or $p_2 = p_{2\rho}$ 1 is "left" in the state

$$\delta(q_1 - q_{2\mu} + Q) \quad \text{or} \quad \frac{1}{\sqrt{h}} e^{\frac{i}{\hbar} q_1 (P - p_{2\rho})} \qquad (2.99)$$

and $q_1 = q_{2\mu} - Q$ or $p_1 = p - p_{2\rho}$ respectively. In this way the coordinate or momentum of 1 is measured by the coordinate or momentum of 2 after the interaction between 1 and 2. We come back to this example in 5.06.

3. Operator relations.

3.01 *Exponentials.* In the ring of operators **a** generated by two non-commuting **Hermitian** basic operators **p** and **q**, for which

$$[\mathbf{p},\mathbf{q}] = 1, \text{ i.e. } \mathbf{pq} - \mathbf{qp} = \frac{\hbar}{i} (\hbar > 0), \qquad (3.01)$$

we are going to derive a **Fourier** expansion similar to that in a commutative ring of functions $a(p,q)$ of two real basic variables p and q. For this purpose we need some exponential relations. It should be remembered that we still have a rather specialized case, because the commutator (3.01) of **p** and **q** commutes with **p** and **q**.

With (3.01) one has [2]

$$e^{\frac{i}{\hbar} (\mathbf{p}+\mathbf{q})} = \lim_{n\to\infty}\left(1 + \frac{1}{n}\frac{i}{\hbar}(\mathbf{p}+\mathbf{q})\right)^n = \lim_{n\to\infty}\left((1 + \frac{1}{n}\frac{i}{\hbar}\mathbf{p})(1 + \frac{1}{n}\frac{i}{\hbar}\mathbf{q})\right)^n$$

$$= \lim_{n\to\infty}(1 + \frac{1}{n}\frac{i}{\hbar}\mathbf{p})^n(1 + \frac{1}{n}\frac{i}{\hbar}\mathbf{q})^n\left(1 - \frac{1}{n^2}\frac{i}{\hbar}\right)^{\frac{(n-1)n}{2}} =$$

$$= e^{\frac{i}{\hbar}\mathbf{p}} e^{\frac{i}{\hbar}\mathbf{q}} e^{-\frac{i}{2\hbar}}. \qquad (3.02)$$

With $(x\mathbf{p} + y\mathbf{q})$ and $(x'\mathbf{p} + y'\mathbf{q})$ instead of $\mathbf{p}$ and $\mathbf{q}$ we get for (3.01)

$$[(x\mathbf{p} + y\mathbf{q}), (x'\mathbf{p} + y'\mathbf{q})] = xy' - yx' \qquad (3.03)$$

and for (3.02)

$$e^{\frac{i}{\hbar}((x+x')\mathbf{p} + (y+y')\mathbf{q})} = e^{\frac{i}{\hbar}(x\mathbf{p}+y\mathbf{q})}\, e^{\frac{i}{\hbar}(x'\mathbf{p}+y'\mathbf{q})}\, e^{-\frac{i}{2\hbar}(xy'-yx')}. \qquad (3.04)$$

(Important special cases are $y = x' = 0$ or $x = y' = 0$). Further

$$e^{-\frac{i}{\hbar}(\xi\mathbf{p}+\eta\mathbf{q})}\, e^{\frac{i}{\hbar}(x\mathbf{p}+y\mathbf{q})}\, e^{\frac{i}{\hbar}(\xi\mathbf{p}+\eta\mathbf{q})} = e^{\frac{i}{\hbar}(x\mathbf{p}+y\mathbf{q})}\, e^{\frac{i}{\hbar}(x\eta-y\xi)}. \qquad (3.05)$$

Analogous to the (symbolical) relation

$$\frac{1}{h^2}\iint dq\, dq\, e^{\frac{i}{\hbar}(xp+yq)} = \delta(x)\,\delta(y), \qquad (3.06)$$

(3.05) gives the operator relation

$$\frac{1}{h^2}\iint d\xi\, d\eta\, e^{-\frac{i}{\hbar}(\xi\mathbf{p}+\eta\mathbf{q})}\, e^{\frac{i}{\hbar}(x\mathbf{p}+y\mathbf{q})}\, e^{\frac{i}{\hbar}(\xi\mathbf{p}+\eta\mathbf{q})} = \delta(x)\,\delta(y). \qquad (3.07)$$

Further analogous to

$$\frac{1}{h^2}\iiiint dx\, dy\, dp'\, dq'a(p',q')\, e^{-\frac{i}{\hbar}(xp'+yq')}\, e^{\frac{i}{\hbar}(xp+yq)} = a(p,q), \qquad (3.08)$$

we have

$$\frac{1}{h^2}\iiiint dx\, dy\, d\xi\, d\eta\, e^{-\frac{i}{\hbar}(\xi\mathbf{p}+\eta\mathbf{q})}\, \mathbf{a}\, e^{-\frac{i}{\hbar}(x\mathbf{p}+y\mathbf{q})}\, e^{\frac{i}{\hbar}(\xi\mathbf{p}+\eta\mathbf{q})}\, e^{\frac{i}{\hbar}(x\mathbf{p}+y\mathbf{q})}$$

$$= \frac{1}{h^2}\iiiint dx\, dy\, d\xi\, d\eta\, e^{-\frac{i}{\hbar}(\xi\mathbf{p}+\eta\mathbf{q})}\, \mathbf{a}\, e^{\frac{i}{\hbar}(\xi\mathbf{p}+\eta\mathbf{q})}\, e^{\frac{i}{\hbar}(\xi y-\eta x)}$$

$$= \iint d\xi\, d\eta\, e^{-\frac{i}{\hbar}(\xi\mathbf{p}+\eta\mathbf{q})}\, \mathbf{a}\, e^{\frac{i}{\hbar}(\xi\mathbf{p}+\eta\mathbf{q})}\, \delta(\xi)\,\delta(\eta) = \mathbf{a}. \qquad (3.09)$$

In the same way as (3.08) and (3.06) show that every (normalizable) function $a(p,q)$ can be expanded into a F o u r i e r integral

$$a(p,q) = \iint dx\, dy\, \alpha(x,y)\, e^{\frac{i}{\hbar}(xp+yq)}$$

with $\qquad\qquad\qquad\qquad\qquad\qquad\qquad\qquad\qquad\qquad$ (3.10)

$$\alpha(x,y) = \frac{1}{h^2}\iint dp\, dq\, a(p,q)\, e^{-\frac{i}{\hbar}(xp+yq)},$$

(3.09) and (3.07) show that every operator $\mathbf{a}$ (with adjoint $\mathbf{a}^\dagger$) can be expanded into

$$\mathbf{a} = \iint dx\, dy\, \alpha(x,y)\, e^{\frac{i}{\hbar}(x\mathbf{p}+y\mathbf{q})}$$

with $\alpha(x,y) = \dfrac{1}{h^2} \displaystyle\iint d\xi\, d\eta\, e^{-\frac{i}{\hbar}(\xi p + \eta q)}\, \mathbf{a}\, e^{-\frac{i}{\hbar}(x p + y q)}\, e^{\frac{i}{\hbar}(\xi p + \eta q)}$. (3.11)

This is already the F o u r i e r expansion, but the coefficients $\alpha(x,y)$ can still be expressed in a more simple form.

3.02 *The trace.* When $\mathbf{U}$ is a unitary operator

$$\mathbf{U}^\dagger \mathbf{U} = 1, \tag{3.12}$$

the unitary transformation

$$\mathbf{a}' = \mathbf{U}^\dagger \mathbf{a} \mathbf{U}; \quad \varphi' = \mathbf{U}\varphi, \quad \varphi'^\dagger = \varphi^\dagger \mathbf{U}^\dagger \tag{3.13}$$

leaves all operator relations invariant. Therefore the latter can be derived in a suitably chosen representation.

The eigenvalues q of $\mathbf{q}$ and p of $\mathbf{p}$ are assumed to run continuously between $-\infty$ and $+\infty$. In q-representation the operators $\mathbf{q}$ and $\mathbf{p}$ can be taken in the form

$$\mathbf{q} = \mathbf{q}^\dagger = q,\, \mathbf{p} = \mathbf{p}^\dagger = \frac{\hbar}{i}\frac{\partial}{\partial q} \quad \text{or} \quad -\frac{\hbar}{i}\frac{\delta}{\delta q} \tag{3.14}$$

($\delta/\delta q$ is meant to operate to the left). With (3.04) we can write

$$e^{\frac{i}{\hbar}(x p + y q)} = e^{\frac{i}{2\hbar}x\mathbf{p}}\, e^{\frac{i}{\hbar}y q}\, e^{\frac{i}{2\hbar}x\mathbf{p}} = e^{\frac{x}{2}\frac{\partial}{\partial q}}\, e^{\frac{i}{\hbar}y q}\, e^{\frac{x}{2}\frac{\partial}{\partial q}}. \tag{3.15}$$

Expressing occasionally the inner product explicitly by an integral, we get with (1.09), (3.15) and (1.05)

$$\frac{1}{h} Tr\, e^{\frac{i}{\hbar}(x p + y q)} = \frac{1}{h} \sum_\mu \int dq\, \varphi_\mu^\dagger(q)\, e^{-\frac{x}{2}\frac{\delta}{\delta q}}\, e^{\frac{i}{\hbar}y q}\, e^{\frac{x}{2}\frac{\partial}{\partial q}}\, \varphi_\mu(q)$$

$$= \frac{1}{h} \sum_\mu \int dq\, \varphi_\mu^\dagger\!\left(q - \frac{x}{2}\right) e^{\frac{i}{\hbar}y q}\, \varphi_\mu\!\left(q + \frac{x}{2}\right) = \delta(x)\, \delta(y). \tag{3.16}$$

The result is independent of the chosen representation. Comparing (3.16) with (3.07) and remembering the linear expansion (3.11) of $\mathbf{a}$, we see that $Tr\mathbf{a}$ can invariantly be represented by the operator relation

$$\frac{1}{h} Tr\mathbf{a} = \frac{1}{h^2} \iint d\xi\, d\eta\, e^{-\frac{i}{\hbar}(\xi p + \eta q)}\, \mathbf{a}\, e^{\frac{i}{\hbar}(\xi p + \eta q)}. \tag{3.17}$$

3.03 *F o u r i e r expansion.* Rewriting (3.07), (3.09) and (3.11) with the help of (3.17) we get

$$\frac{1}{h} Tr\, e^{\frac{i}{\hbar}(x p + y q)} = \delta(x)\, \delta(y), \tag{3.18}$$

$$\frac{1}{h} \iint dx\, dy\, Tr\left(\mathbf{a}\, e^{-\frac{i}{\hbar}(x\mathbf{p}+y\mathbf{q})}\right) e^{\frac{i}{\hbar}(x\mathbf{p}+y\mathbf{q})} = \mathbf{a} \qquad (3.19)$$

and

$$\mathbf{a} = \iint dx\, dy\, \alpha(x,y)\, e^{\frac{i}{\hbar}(x\mathbf{p}+y\mathbf{q})} \cdot$$

with $\qquad\qquad\qquad\qquad\qquad\qquad\qquad\qquad\qquad$ (3.20)

$$\alpha(x,y) = \frac{1}{h} Tr\left(\mathbf{a}\, e^{-\frac{i}{\hbar}(x\mathbf{p}+y\mathbf{q})}\right).$$

(3.18), (3.19) and (3.20) are entirely analogous to (1.13), (1.14) and (1.15). (3.18) and (3.19) respectively express the orthonormality and the completeness of the systems of operators

$$\frac{1}{\sqrt{h}}\, e^{\frac{i}{\hbar}(x\mathbf{p}+y\mathbf{q})} \quad \text{(with variable } x \text{ and } y\text{).}$$

(1.15) and (3.20) are the two ways we use for the expansions of operators.

4. Correspondence.

4.01 *v o n N e u m a n n's rules.* We now examine the rules of correspondence I, II, III, IV and V'. First I and II.

We show that if between the elements a of one ring and the elements $\mathbf{a}$ of another ring there is a one-to-one correspondence $a \longleftrightarrow \mathbf{a}$, which satisfies v o n N e u m a n n's rules (cf. 1.10)

$$\text{if } a \longleftrightarrow \mathbf{a}, \text{ then } f(a) \longleftrightarrow f(\mathbf{a}), \qquad\qquad \text{I}$$

$$\text{if } a \longleftrightarrow \mathbf{a} \text{ and } b \longleftrightarrow \mathbf{b}, \text{ then } a + b \longleftrightarrow \mathbf{a} + \mathbf{b}, \qquad \text{II}$$

the two rings are isomorphous.

We get using I and II

$$(a + b)^2 - a^2 - b^2 = ab + ba \longleftrightarrow \mathbf{ab} + \mathbf{ba} \qquad (4.01)$$

and also using (4.01)

$$a(ab + ba) + (ab + ba)a - a^2b - ba^2 = 2aba \longleftrightarrow 2\mathbf{aba} \qquad (4.02)$$

and further using (4.02)

$$(ab + ba)^2 - b(2aba) - (2aba)b =$$

$$= -(ab - ba)^2 \longleftrightarrow -(\mathbf{ab} - \mathbf{ba})^2. \qquad (4.03)$$

Therefore we have

$$ab - ba \longleftrightarrow \pm (\mathbf{ab} - \mathbf{ba}), \qquad (4.04)$$

and with (4.01)

$ab \longleftrightarrow \mathbf{ab}$ (for all a and b) or $ab \longleftrightarrow \mathbf{ba}$ (for all a and b). (4.05)

This means that the rings are isomorphous.

It follows that, if one ring is commutative and the other not, I and II are inconsistent [9]. (When the commutators are of the order of $\hbar$, the discrepancy is according to (4.03) of the order of $\hbar^2$).

4.02 *Bracket expressions.* Then V'. For the correspondence $a \longleftrightarrow \mathbf{a}$ between the commutative ring with generating elements p and q and the non-commutative ring with generating elements $\mathbf{p}$ and $\mathbf{q}$ with commutator (3.01) ($p \longleftrightarrow \mathbf{p}$ and $q \longleftrightarrow \mathbf{q}$) we show that the rule (cf. 1.18)

if $a(p,q) \longleftrightarrow \mathbf{a}$ and $b(p,q) \longleftrightarrow \mathbf{b}$, then $(a(p,q), b(p,q)) \longleftrightarrow [\mathbf{a},\mathbf{b}]$ V'

is self contradictory.

With

$$p^2 \longleftrightarrow \mathbf{x}_1, \; q^2 \longleftrightarrow \mathbf{x}_2; \; p^3 \longleftrightarrow \mathbf{y}_1, \; q^3 \longleftrightarrow \mathbf{y}_2 \qquad (4.06)$$

we find from

$$\begin{aligned}
\tfrac{1}{2}(p^2,q) = p &\longleftrightarrow \tfrac{1}{2}[\mathbf{x}_1,\mathbf{q}] = \mathbf{p}, \\
\tfrac{1}{2}(p^2,p) = 0 &\longleftrightarrow \tfrac{1}{2}[\mathbf{x}_1,\mathbf{p}] = 0
\end{aligned} \qquad (4.07)$$

(and similar relations for q^2 and $\mathbf{x}_2$) that

$$p^2 \longleftrightarrow \mathbf{p}^2 + c_1, \; q^2 \longleftrightarrow \mathbf{q}^2 + c_2 \qquad (4.08)$$

and from

$$\begin{aligned}
\tfrac{1}{3}(p^3,q) = p^2 &\longleftrightarrow \tfrac{1}{3}[\mathbf{y}_1,\mathbf{q}] = \mathbf{p}^2 + c_1, \\
\tfrac{1}{3}(p^3,p) = 0 &\longleftrightarrow \tfrac{1}{3}[\mathbf{y}_1,\mathbf{p}] = 0
\end{aligned} \qquad (4.09)$$

(and similar relations for q^3 and $\mathbf{y}_2$) that

$$p^3 \longleftrightarrow \mathbf{p}^3 + 3c_1\mathbf{p} + d_1, \; q^3 \longleftrightarrow \mathbf{q}^3 + 3c_2\mathbf{q} + d_2 \quad (4.10)$$

($c_1, c_2; d_1, d_2$ are undetermined constants). Further we get

$$\begin{aligned}
\tfrac{1}{6}(p^3,q^2) = p^2 q &\longleftrightarrow \tfrac{1}{6}[(\mathbf{p}^3 + 3c_1\mathbf{p} + d_1),(\mathbf{q}^2 + c_2)] = \tfrac{1}{2}(\mathbf{p}^2\mathbf{q} + \mathbf{q}\mathbf{p}^2) + c_1\mathbf{q}, \\
pq^2 &\longleftrightarrow \qquad\qquad\qquad\qquad\qquad\qquad \tfrac{1}{2}(\mathbf{p}\mathbf{q}^2 + \mathbf{q}^2\mathbf{p}) + c_2\mathbf{p}
\end{aligned} \qquad (4.11)$$

and

$$\begin{aligned}
\tfrac{1}{9}(p^3,q^3) = p^2 q^2 &\longleftrightarrow \tfrac{1}{9}[(\mathbf{p}^3 + 3c_1\mathbf{p} + d_1), (\mathbf{q}^3 + 3c_2\mathbf{q} + d_2)] \\
&= \tfrac{1}{2}(\mathbf{p}^2\mathbf{q}^2 + \mathbf{q}^2\mathbf{p}^2) + \tfrac{1}{3}h^2 + c_1\mathbf{q}^2 + c_2\mathbf{p}^2 + c_1c_2. \quad (4.12)
\end{aligned}$$

With (4.11) we get

$$\tfrac{1}{3}(p^2q,pq^2) = p^2q^2 \longleftrightarrow \tfrac{1}{3}[(\tfrac{1}{2}(\mathbf{p^2q+qp^2}) + c_1\mathbf{q}), (\tfrac{1}{2}(\mathbf{pq^2+q^2p}) + c_2\mathbf{p})]$$
$$= \tfrac{1}{2}(\mathbf{p^2q^2+q^2p^2}) + \tfrac{2}{3}\hbar^2 - c_1\mathbf{q}^2 - c_2\mathbf{p}^2 - \tfrac{1}{3}c_1c_2. \quad (4.13)$$

(4.12) and (4.13) can only be identical for $c_1 = c_2 = 0$ and $\hbar = 0$. Therefore V' is self inconsistent (the deficiency is of the order of $\hbar^2$).

4.03 *W e y l's correspondence*. And finally III and IV with parameters p and q (i.e. for the same rings as in 4.02). We denote the density function by $\rho(p,q)$. The rules (cf. 1.13)

$$1 \longleftrightarrow \mathbf{1}, \qquad\qquad\qquad \text{III}$$

if $a(p,q) \longleftrightarrow \mathbf{a}$ and $b(p,q) \longleftrightarrow \mathbf{b}$,

$$\text{then } \iint dp\, dq\, \rho(p,q)\, a(p,q)\, b(p,q) = Tr(\mathbf{ab}) \quad \text{IV}$$

can be satisfied by (1.55)

$$a(p,q) = Tr(\mathbf{m}(p,q)\mathbf{a}), \quad \mathbf{a} = \iint dp\, dq\, \rho(p,q)\, \mathbf{m}(p,q)\, a(p,q) \quad (4.14)$$

with a transformation nucleus $\mathbf{m}(p,q)$, which satisfies (1.57), (1.58); (1.59), (1.60)

$$Tr\,\mathbf{m}(p,q) = 1, \qquad\qquad (4.15)$$

$$\iint dp\, dq\, \rho(p,q)\, \mathbf{m}(p,q) = \mathbf{1}; \qquad\qquad (4.16)$$

$$Tr(\mathbf{m}(p,q)\, \mathbf{m}(p',q')) = \rho^{-1}(p,q)\, \delta(p-p')\, \delta(q-q'), \quad (4.17)$$

$$\iint dp\, dq\, \rho(p,q)\, Tr(\mathbf{m}(p,q)\, \mathbf{a})\, Tr(\mathbf{m}(p,q)\, \mathbf{b}) =$$
$$= Tr(\mathbf{ab}) \text{ (for every } \mathbf{a} \text{ and } \mathbf{b}), \quad (4.18)$$

When we replace in (1.56) the complete orthonormal systems $k^*_{\mu\nu}(p,q)$ of (1.54) and $\mathbf{k}_{\mu\nu}$ of (1.15) by the complete orthonormal systems

$$\frac{1}{h} e^{-\frac{i}{\hbar}(xp+yq)} \text{ of (3.10) and } e^{\frac{i}{\hbar}(\mathbf{xp+yq})} \text{ of (3.20)},$$

we find a solution

$$\mathbf{m}(p,q) = \frac{1}{h}\iint dx\, dy\, e^{\frac{i}{\hbar}(\mathbf{xp+yq})}\, e^{-\frac{i}{\hbar}(xp+yq)} \quad (4.19)$$

of (4.15), (4.16); (4.17), (4.18) with the density function

$$\rho(p,q) = \frac{1}{h}. \qquad\qquad (4.20)$$

Then we get for (4.14)

$$a(p,q) = \frac{1}{h}\iint dx\, dy\, e^{\frac{i}{\hbar}(xp+yq)}\, Tr\left(e^{-\frac{i}{\hbar}(\mathbf{xp+yq})}\, \mathbf{a}\right),$$
$$\mathbf{a} = \frac{1}{h}\iint dx\, dy\, e^{\frac{i}{\hbar}(\mathbf{xp+yq})}\, \frac{1}{h}\iint dp\, dq\, e^{-\frac{i}{\hbar}(xp+yq)}\, a(p,q).$$
$$(4.21)$$

With the Fourier expansions (3.10) and (3.20) this correspondence reads

$$\iint dx\, dy\, \alpha(x,y)\, e^{\frac{i}{\hbar}(xp+yq)} \longleftrightarrow \iint dx\, dy\, \alpha(x,y)\, e^{\frac{i}{\hbar}(xp+yq)}, \quad (4.22)$$

which is Weyl's correspondence [2]).

II is a consequence of IV and is therefore satisfied by the correspondence (4.21). We will see what is left of I and V'. If $a \longleftrightarrow \mathbf{a}$ and $b \longleftrightarrow \mathbf{b}$ according to (4.21) we find with (3.04)

$$\mathbf{ab} = \frac{1}{h^4} \iint \cdots \iint dx\, dy\, dx'\, dy'\, dp\, dq\, dp'\, dq' \;.$$

$$\cdot\, e^{\frac{i}{\hbar}((x+x')\mathbf{p}+(y+y')\mathbf{q})}\, e^{\frac{i}{2\hbar}(xy'-yx')}\, e^{-\frac{i}{\hbar}(xp+yq+x'p'+y'q')}\, a(p,q)\, b(p,q). \quad (4.23)$$

With the variables

$$\begin{aligned}
\xi &= x+x', & \eta &= y+y', & \sigma &= \frac{p+p'}{2}, & \tau &= \frac{q+q'}{2}, \\
\xi' &= \frac{x-x'}{2}, & \eta' &= \frac{y-y'}{2}, & \sigma' &= p-p', & \tau' &= q-q',
\end{aligned} \quad (4.24)$$

this becomes

$$\mathbf{ab} = \frac{1}{h^4} \iint \cdots \iint d\xi\, d\eta\, d\xi'\, d\eta'\, d\sigma\, d\tau\, d\sigma'\, d\tau'\, e^{\frac{i}{\hbar}(\xi\mathbf{p}+\eta\mathbf{q})}\, e^{\frac{i}{2\hbar}(-\xi\eta'+\eta\xi')}$$

$$e^{-\frac{i}{\hbar}(\xi\sigma+\eta\tau+\xi'\sigma'+\eta'\tau')}\, a(\sigma+\tfrac{1}{2}\sigma',\, \tau-\tfrac{1}{2}\tau')\, b(\sigma-\tfrac{1}{2}\sigma',\, \tau+\tfrac{1}{2}\tau')$$

$$= \frac{1}{h^2} \iiiint d\xi\, d\eta\, d\sigma\, d\tau\, e^{\frac{i}{\hbar}(\xi\mathbf{p}+\eta\mathbf{q})}\, e^{-\frac{i}{\hbar}(\xi\sigma+\eta\tau)}$$

$$a(\sigma+\tfrac{1}{4}\eta,\, \tau-\tfrac{1}{4}\xi)\, b(\sigma-\tfrac{1}{4}\eta,\, \tau+\tfrac{1}{4}\xi)$$

$$= \frac{1}{h^2} \iiiint d\xi\, d\eta\, d\sigma\, d\tau\, e^{\frac{i}{\hbar}(\xi\mathbf{p}+\eta\mathbf{q})}\, e^{-\frac{i}{\hbar}(\xi\sigma+\eta\tau)}$$

$$\left(e^{\frac{1}{4}(\eta\frac{\partial}{\partial\sigma}-\xi\frac{\partial}{\partial\tau})} a(\sigma,\tau) \right) \left(e^{-\frac{1}{4}(\eta\frac{\partial}{\partial\sigma}-\xi\frac{\partial}{\partial\tau})} b(\sigma,\tau) \right). \quad (4.25)$$

The expressions in brackets at the end are a symbolical representation of Taylor expansion. With the substitution

$$\xi \to x, \eta \to y, \sigma \to p, \tau \to q \quad (4.26)$$

we get by partial integration

$$\mathbf{ab} = \frac{1}{h} \iint dx\, dy\, e^{\frac{i}{\hbar}(x\mathbf{p}+y\mathbf{q})} \frac{1}{h} \iint dp\, dq \;.$$

$$\cdot\, e^{-\frac{i}{\hbar}(xp+yq)} \left(a(p,q)\, e^{\frac{\hbar}{2i}(\frac{\partial}{\partial p}\frac{\partial}{\partial q}-\frac{\partial}{\partial q}\frac{\partial}{\partial p})} b(p,q) \right). \quad (4.27)$$

This gives for the H e r m i t i a n operators $\frac{1}{2}$ (**ab** + **ba**) and $\frac{i}{2}$.
(**ab** — **ba**) the correspondence

$$a(p,q) \cos \frac{\hbar}{2} \left(\frac{\delta}{\delta p} \frac{\partial}{\partial q} - \frac{\delta}{\delta q} \frac{\partial}{\partial p} \right) b(p,q) \longleftrightarrow \tfrac{1}{2} (\mathbf{ab} + \mathbf{ba}), \quad (4.28)$$

$$a(p,q) \sin \frac{\hbar}{2} \left(\frac{\delta}{\delta p} \frac{\partial}{\partial q} - \frac{\delta}{\delta q} \frac{\partial}{\partial p} \right) b(p,q) \longleftrightarrow \frac{i}{2} (\mathbf{ab} - \mathbf{ba}). \quad (4.29)$$

To the neglect of terms of order of $\hbar^2$ and higher (4.28) and (4.29)
would read

$$a(p,q) \, b(p,q) \longleftrightarrow \tfrac{1}{2} (\mathbf{ab} + \mathbf{ba}), \quad (4.30)$$

$$a(p,q) \frac{\hbar}{2} \left(\frac{\delta}{\delta p} \frac{\partial}{\partial q} - \frac{\delta}{\partial q} \frac{\partial}{\partial p} \right) b(p,q) \longleftrightarrow \frac{i}{2} (\mathbf{ab} - \mathbf{ba}). \quad (4.31)$$

(4.30) would lead to I, (4.31) is equivalent to V'.

We examine which functions $f(a)$ satisfy I. From (4.28) we see
that the correspondence

$$\text{if } a \longleftrightarrow \mathbf{a}, \text{ then } a^n \longleftrightarrow \mathbf{a}^n \text{ (for every integer } n) \quad (4.32)$$

only holds if

$$a^k \cos \frac{\hbar}{2} \left(\frac{\delta}{\partial p} \frac{\partial}{\partial q} - \frac{\delta}{\delta q} \frac{\partial}{\partial p} \right) a^l = a^{k+l} \text{ (for all integers } k \text{ and } l). \quad (4.33)$$

First take for a a homogeneous polynomial in p and q of degree n.
An elementary calculation shows that the condition

$$a \cos \frac{\hbar}{2} \left(\frac{\delta}{\delta p} \frac{\partial}{\partial q} - \frac{\delta}{\delta q} \frac{\partial}{\partial p} \right) a = a^2 \quad (4.34)$$

or

$$a \left(\frac{\delta}{\delta p} \frac{\partial}{\partial q} - \frac{\delta}{\delta q} \frac{\partial}{\partial p} \right)^{2k} a = a^2 \text{ (for } 0 < 2k \leqslant n) \quad (4.35)$$

is only satisfied if a is of the form $(xp + yq)^n$. Then it follows that
any polynomial in p and q can only satisfy (4.33) if it is a polynomial in $xp + yq$. This finally means that I can only be satisfied
if a is a function of a certain linear combination $xp + yq$ of p and q.
With the help of the F o u r i e r expansion (4.22) it is easily seen that
every (normalizable) function of $xp + yq$ does satisfy I. Therefore the
least restricted form of I, which is consistent with the correspondence
(4.21) is

$$f(xp + yq) \longleftrightarrow f(x\mathbf{p} + y\mathbf{q}). \quad (4.36)$$

As to V', we see from (4.31) that for the correspondence (4.21) the bracket expression $((a(p,q), b(p,q)))$ (cf. 1.14) defined by

if $a(p,q) \longleftrightarrow \mathbf{a}$ and $b(p,q) \longleftrightarrow \mathbf{b}$, then $((a(p,q), b(p,q))) \longleftrightarrow [\mathbf{a},\mathbf{b}]$ (4.37)

is given by

$$((a(p,q),b(p,q))) = a(p,q) \frac{2}{\hbar} \sin\left(\frac{\hbar}{2} \frac{\delta}{\delta p} \frac{\partial}{\partial q} - \frac{\delta}{\delta q} \frac{\partial}{\partial p}\right) b(p,q). \quad (4.38)$$

If $a(p,q)$ or $b(p,q)$ is a polynomial in p and q of at most 2nd degree, we have a special case for which the bracket expressions $((a,b))$ and (a,b) coincide.

The correspondence (4.21) is a solution of III and IV. We have not investigated the possibility of other solutions with the same parameters p and q.

5. Quasi-distributions.

5.01 *Proper and improper representations.* With Weyl's correspondence (4.22) as a special solution of

$$\mathbf{1} \longleftrightarrow 1 \qquad\qquad\qquad \text{III}$$

if $\mathbf{k} \longleftrightarrow k(p,q)$ and $\mathbf{a} \longleftrightarrow a(p,q)$,

$$\text{then } Tr(\mathbf{ka}) = \frac{1}{h} \iint dp\, dq\, k(p,q)\, a(p,q) \qquad\qquad \text{IV}$$

(with parameters p and q and density function $\rho(p,q) = 1/h$), we obtain a special case of a transformation between a representation in terms of operators $\mathbf{k}$ and $\mathbf{a}$ and a representation in terms of functions $k(p,q)$ and $a(p,q)$. Quantum statistics are usually represented in terms of operators, classical statistics in terms of functions. We assert that the usual description is also the proper one. The statistical operator $\mathbf{k}$ of the quantum representation and the statistical distribution function $k(p,q)$ of the classical representation are non-negative definite, but in general the quantum $k(p,q)$ and the classical $\mathbf{k}$ are not. This makes that for orthogonal states, for which

$$Tr(\mathbf{k_1 k_2}) = \frac{1}{h} \iint dp\, dq\, k_1(p,q)\, k_2(p,q) = 0, \qquad (5.01)$$

the product $\mathbf{k_1 k_2}$ or $k_1(p,q)k_2(p,q)$ vanishes in the proper representation, but in the improper representation it need not. The equations

4

of motion of the quantum **k** are described by infinitesimal unitary transformations, those of the classical $k(p,q)$ by infinitesimal canonical transformations (contact transformations), but the equations of motion of the classical **k** and the quantum $k(p,q)$ are in general not of these types. Because the improper representation is formally equivalent to the proper one, it is (provided it is not misinterpreted) a correct description, though it is in general a rather impracticable one.

In spite of its deficiences, or rather because of them, we discuss some aspects of the improper representation of quantum mechanics in terms of $k(p,q)$ and $a(p,q)$, i.e. the quasi-statistical description of the 1st kind Q^1 (cf. 1.19). It more or less illustrates the ways along which some opponents might hope to escape B o h r's reasonings and v o n N e u m a n n's proof and the places where they are dangerously near breaking their necks.

5.02 *Transition functions.* For the transition functions $k_{\mu\nu}(p,q)$ corresponding to the transition operators (1.03) according to (4.21) we find with the help of the q-representation (occasionally expressing the inner product explicitly by an integral) similar to (3.16)

$$k_{\mu\nu}(p,q) = \frac{1}{h}\int\!\!\int dx\, dy\, e^{\frac{i}{\hbar}(xp+yq)}\int dq'\, \varphi^\dagger_\mu(q')\, e^{\frac{x}{2}\frac{\delta}{\delta q'}}\, e^{\frac{i}{\hbar}yq}\, e^{-\frac{x}{2}\frac{\partial}{\partial q'}}\, \varphi_\nu(q')$$

$$= \int dx\, \varphi^\dagger_\mu(q)\, e^{\frac{x}{2}\frac{\delta}{\delta q}}\, e^{\frac{i}{\hbar}xp}\, e^{-\frac{x}{2}\frac{\partial}{\partial q}}\, \varphi_\nu(q)$$

$$= \int dx\, \varphi^\dagger_\mu\!\left(q + \frac{x}{2}\right) e^{\frac{i}{\hbar}xp}\, \varphi_\nu\!\left(q - \frac{x}{2}\right). \tag{5.02}$$

Because the wave functions φ_μ are only determined but for a factor $e^{i/\hbar\,\gamma_\mu}$ (γ real), the $k_{\mu\nu}(p,q)$ are only determined but for a factor $e^{i/\hbar\,(\gamma_\mu-\gamma_\nu)}$. The distribution functions, which are thus obtained with W e y l's correspondence [2]) become identical to those given by W i g n e r [10]).

5.03 *Proper value.* In a distribution **k** or $k(p,q)$ a quantity **a** or $a(p,q)$ can be regarded to have a proper value if the condition (2.10)

$$Tr(\mathbf{k}f(a)) = f(Tr(\mathbf{k}a)) \tag{5.03}$$

or

$$\frac{1}{h}\int\!\!\int dp\, dq\, k(p,q)\, f(a(p,q)) = f\left(\frac{1}{h}\int\!\!\int dp\, dq\, k(p,q)\, a(p,q)\right) \tag{5.04}$$

is satisfied for every f. Whereas the validity of (5.04) is for a proper (non-negative definite) $k(p,q)$ already guaranteed by the validity of the special case $f(a) = a^2$, it is not for a proper $\mathbf{k}$ or an improper $k(p,q)$. For a proper $\mathbf{k}$ the validity of (5.03) or (2.11) requires that $\mathbf{a}$ is of the form

$$a(x\mathbf{p} + y\mathbf{q}) \tag{5.05}$$

and $\mathbf{k}$ an eigenstate of $\mathbf{a}$. For any $k(p,q)$ the validity of (5.04) requires that $k(p,q)$ is of the form

$$\delta(a(p,q) - a_\mu), \tag{5.06}$$

which is a proper (i.e. non-negative definite) one. Because (5.03) and (5.04) are identical, the conditions (5.05) and (5.06) are equivalent. This means that the eigenstates of the operators $a(x\mathbf{p} + y\mathbf{q})$ and of no other operators correspond with proper (and orthonormal and therefore non-overlapping) distributions of the form (5.06), in which a_μ is the corresponding eigenvalue. This case would be rather encouraging for a statistical description of the 1st kind S^1, if it were not just an exceptional case.

The eigenfunctions of $a(x\mathbf{p} + y\mathbf{q})$ are in q-representation

$$\varphi_\rho(q) = \frac{1}{\sqrt{xh}} e^{\frac{i}{\hbar}\left(-\frac{1}{2xy}(yq-\rho)^2 + \gamma(\rho)\right)} \quad \text{for } x \neq 0,$$

$$\varphi_\rho(q) = \sqrt{y}\, \delta(yq - \rho)\, e^{\frac{i}{\hbar}\gamma(\rho)} \qquad \text{for } x = 0. \tag{5.07}$$

($\gamma(\rho)$ real arbitrary). The corresponding eigenvalues are $a(\rho)$

$$a(x\mathbf{p} + y\mathbf{q})\varphi_\rho = a(\rho)\varphi_\rho. \tag{5.08}$$

ρ, which is the eigenvalue of $x\mathbf{p} + y\mathbf{q}$ (for arbitrary fixed x and y), runs between $-\infty$ and $+\infty$. The domain of eigenvalues of $a(x\mathbf{p} + y\mathbf{q})$ is therefore the same as that of the functions $a(z)$ ($-\infty \leqslant z \leqslant \infty$). This means that the domain of the proper values of observables, which have such, are unrestricted by quantum conditions.

Inserting the eigenfunctions (5.07) in (5.02) we get

$$k_{\mu\nu}(p,q) = \delta\left(xp + yq - \frac{p_\mu + p_\nu}{2}\right) e^{-\frac{i}{\hbar}\left(\left(\frac{p}{y} - \frac{q}{x}\right)\frac{p_\mu - p_\nu}{2} + \gamma'(p_\mu) - \gamma'(p_\nu)\right)}. \tag{5.09}$$

(The expression in brackets in the exponent in (5.09) is a canonical conjugate of $xp + yq$). The $k_{\mu\mu}(p,q)$ are actually of the form (5.06).

5.04 *The harmonic oscillator*. After we have treated in 5.03 a special case for which the $k(p,q)$ are of proper type themselves, we now deal with a case for which their equations of motion are of proper type. According to (1.43) and condition V′ they are if $((H(p,q),k(p,q)))$ coincides with $(H(p,q),k(p,q))$ and according to (4.38) this is the case for every $k(p,q)$ if $H(p,q)$ is a polynomial in p and q of at most 2nd degree. This condition is satisfied for the harmonic oscillator, for which $H(p,q)$ coincides with the classical Hamiltonian

$$H(p,q) \doteq \frac{p^2}{2m} + \frac{m\omega^2}{2} q^2 = \frac{\omega}{2} (p'^2+q'^2); \; p' = \frac{p}{\sqrt{m\omega}}, \; q' = q\sqrt{m\omega}. \quad (5.10)$$

m is the mass, ω the classical circular frequency of the binding. We consider p' and q' as new canonical coordinates and omit the dash.

In q-representation the normalized stationary solutions of the wave equation

$$-\frac{\hbar}{i} \frac{\partial}{\partial t} \varphi_n(q) = \frac{\omega}{2} \left(-\hbar^2 \frac{\partial^2}{\partial q^2} + q^2 \right) \varphi_n(q) \quad (5.11)$$

are

$$\varphi_n(q) = \frac{1}{\sqrt{2^n n!} \, \sqrt{\pi\hbar}} e^{-\frac{1}{2\hbar} q^2} H_n\left(\frac{q}{\sqrt{\hbar}}\right) e^{-in\omega t} \quad (n = 0,1,2,\ldots). \quad (5.12)$$

The Hermitian polynomials $H_n\left(\frac{q}{\sqrt{\hbar}}\right)$ have the generating function

$$e^{\frac{-\xi^2+2\xi q}{\hbar}} = \sum_{n=0}^{\infty} \frac{1}{n!} \left(\frac{\xi}{\sqrt{\hbar}}\right)^n H_n\left(\frac{q}{\sqrt{\hbar}}\right). \quad (5.13)$$

(5.02) becomes with (5.12)

$$k_{mn}(p,q) = \frac{1}{\sqrt{2^{m+n}n!\,m!\pi\hbar}} \int dx \, e^{-\frac{1}{2\hbar}\left(q+\frac{x}{2}\right)^2} H_m\left(\frac{q+\frac{x}{2}}{\sqrt{\hbar}}\right) \cdot$$

$$e^{\frac{i}{\hbar}xp} e^{\frac{1}{2\hbar}\left(q-\frac{x}{2}\right)^2} H_n\left(\frac{q-\frac{x}{2}}{\sqrt{\hbar}}\right) e^{-i(m-n)\omega t}. \quad (5.14)$$

With (5.13) we get

$$\sum_{m,n} \sqrt{\frac{z^{m+n}}{m!n!}} \left(\frac{\xi}{\sqrt{\hbar}}\right)^m \left(\frac{\eta}{\sqrt{\hbar}}\right)^n k_{mn}(p,q)\, e^{i(m-n)\omega t}$$

$$= \frac{1}{\sqrt{\pi\hbar}} \int dx\, e^{\frac{1}{2\hbar}(q+\frac{x}{2})^2 - \frac{1}{\hbar}'\xi - q - \frac{x}{2})^2} e^{\frac{i}{\hbar}xp} e^{\frac{1}{2\hbar}(q-\frac{x}{2})^2 - \frac{1}{\hbar}(\eta - q + \frac{x}{2})^2}$$

$$= 2\, e^{-\frac{1}{\hbar}\left[(q+ip)(q-ip) - 2\xi(q+ip) - 2\eta(q-ip) + 2\xi\eta\right]}$$

$$= 2\, e^{-\frac{1}{\hbar}(q^2+p^2)} \sum_{\mu,\nu,\kappa=0}^{\infty} \frac{1}{\mu!\nu!\kappa!} \left[\frac{2}{\hbar}\xi(q+ip)\right]^{\mu} \left[\frac{2}{\hbar}\eta(q-ip)\right]^{\nu} \left[-\frac{2}{\hbar}\xi\eta\right]^{\kappa}. \quad (5.15)$$

This gives

$$k_{mn}(p,q) = 2\sqrt{m!n!}\, e^{-\frac{1}{\hbar}(p^2+q^2)} \sum_{\kappa=0}^{min\,(m,n)} \frac{(-1)^{\kappa}}{(m-\kappa)!(n-\kappa)!\kappa!} (q+ip)^{m-\kappa}.$$

$$\cdot (q-ip)^{n-\kappa} \left(\frac{2}{\hbar}\right)^{\frac{m+n}{2}-\kappa} e^{-i(m-n)\omega t}$$

$$= 2\sqrt{m!n!}\, e^{-\frac{1}{\hbar}(p^2+q^2)} \sqrt{\frac{2}{\hbar}(p^2+q^2)}^{\,|m-n|} \sum_{\kappa=0}^{min\,(m,n)} \frac{(-1)^{\kappa}}{(m-\kappa)!(n-\kappa)!\kappa!}.$$

$$\cdot \left[\frac{2}{\hbar}(p^2+q^2)\right]^{min\,(m,n)-\kappa} e^{i(m-n)\,arc\,tan\frac{p}{q}} e^{-i(m-n)\omega t}$$

$$= 2(-1)^{max(m,n)} \frac{\sqrt{m!n!}}{max(m,n)!^2} e^{-\frac{1}{2}\left[\frac{2}{\hbar}(p^2+q^2)\right]} \sqrt{\frac{2}{\hbar}(p^2+q^2)}^{\,|m-n|}$$

$$L^{(|m-n|)}_{max(m,n)}\left(\frac{2}{\hbar}(p^2+q^2)\right) e^{i(m-n)\,(arc\,tan\frac{p}{q}-\omega t)}. \quad (5.16)$$

The $L^{(\mu)}_{\lambda}$ are associated L e g e n d r e polynomials. $k_{mn}(p,q)$ is separated into a product of functions of the canonical conjugates $\frac{1}{2}(p^2 + q^2)$ and arc tan (p/q). The $k_{mn}(p,q)$ actually form a complete orthonormal system. For the distribution function $k_{mn}(p,q)$ of the mth eigenstate of $\frac{1}{2}(\mathbf{p}^2 + \mathbf{q}^2)$, the average value of $\frac{1}{2}(p^2 + q^2)$ is $(m+\frac{1}{2})\hbar$, but it is not a proper value.

With (5.10) the transformation (1.47) gives the contact transformation determined by

$$\frac{dp}{dt} = -\omega q, \quad \frac{dq}{dt} = \omega p, \quad (5.17)$$

with solutions

$$p = a\cos(\omega t - \chi), \quad q = a\sin(\omega t - \chi). \quad (5.18)$$

The representative point in the phase space of a superstate rotates uniformly about the origin with constant radius $\sqrt{p^2 + q^2}$ and circular frequency ω. The rotation of the entire distribution $k_{mn}(p,q)$ with this circular frequency ω produces according to the last factor of (5.16) a periodicity with circular frequency $(m - n)\omega$ (like a rotating wheel with $|m - n|$ spokes). Also this would have a hopeful aspect for a description of type S^1, if it were not one out of a few exceptional cases.

5.05 *The scale system*. We shortly return to the measuring process. We start with the most favourable case for a description of the 1st kind S^1 and consider a system l in the measuring chain, for which the distributions $k_{l\mu\mu}(p_l,q_l)$ do not overlap. The corresponding $\mathbf{k}_{l\mu\mu}$ are then eigenstates of an operator of the form $x\mathbf{p}_l + y\mathbf{q}_l$ (cf. 5.03). The scale system is a special case ($x = 0$), which shows all essential features. According to (5.09) we have

$$k_{l\mu\nu}(p_l,q_l) = \delta\left(q_l - \frac{q_{l\mu}+q_{l\nu}}{2}\right) e^{\frac{i}{\hbar}(q_{l\mu}-q_{l\nu})p_l}. \tag{5.19}$$

By ignoration of one or more systems of the measuring chain the non-diagonal functions ($\mu \neq \nu$) are dropped and only the diagonal functions remain. Instead of (5.19) we get

$$k_{l\mu\nu}(p_l,q_l) = \delta(q_l - q_{l\mu})\,\delta(q_{l\mu} - q_{l\nu}). \tag{5.20}$$

(The latter δ-function is actually a remainder of the ignored distribution functions). The effect on (5.19) of ignoration of other systems is formally the same as that of integration over p with density function $1/h$. This illustrates even more plainly than before (cf. 2.07) how the correlation between p_l and other observables is completely destroyed by the reading of q_l. So far there is no difficulty with an interpretation of the 1st kind. We are only concerned with the value of q_l, which is a proper value and uniquely determines the distribution (5.20). The value of p_l is indifferent. As soon as inference is made about other systems in the chain with overlapping $k_{\mu\mu}(p,q)$, correct results are only obtained after the integration over p_l (with density function $1/h$) has been performed (cf. 1.19). In a description of the 1st kind this integration could only be interpreted as an averaging over a great number of measurements. But the integration has already to be performed in a single reading and therefore an interpretation of the 1st kind is excluded.

5.06 *Einstein's paradox*. The multilateral correlated state (2.97) has according to (5.02) the distribution

$$k_{12PQ\ PQ}(p_1,q_1; p_2,q_2) = \delta(q_1 - q_2 + Q)\delta(p_1 + p_2 - P). \quad (5.21)$$

This shows clearly the correlation between q_1 and q_2 and between p_1 and p_2. The similarity to a genuine distribution of the 1st kind is very tempting.

Because (5.21) is highly singular we also consider the distribution

$$k_{12P'Q'P''Q''}(p_1,q_1; p_2,q_2) = \delta\left(q_1 - q_2 + \frac{Q' + Q''}{2}\right)\delta\left(p_1 + p_2 - \frac{P' + P''}{2}\right).$$

$$\cdot e^{-\frac{i}{\hbar}(q_1+q_2)\frac{P'-P''}{2}} e^{\frac{i}{\hbar}(p_1-p_2)\frac{Q'-Q''}{2}} \quad (5.22)$$

(properly instead of (5.21) we should use eigendifferentials). The infringed distribution after a measurement of q_2 or p_2 can be found from (5.22) by integration over p_2 or q_2 respectively with density function $1/h$. This gives

$$\frac{1}{h}\delta\left(q_1 - q_2 + \frac{Q' + Q''}{2}\right) e^{-\frac{i}{\hbar}(q_1+q_2)\frac{P'+P''}{2}} e^{\frac{i}{\hbar}\left(p_1 - \frac{P'+P''}{2}\right)(Q'-Q'')} \quad (5.23)$$

or

$$\frac{1}{h}\delta\left(p_1 + p_2 - \frac{P' + P''}{2}\right) e^{\frac{i}{\hbar}(p_1-p_2)\frac{Q'-Q''}{2}} e^{-\frac{i}{\hbar}\left(q_1+\frac{Q'+Q''}{2}\right)(P'-P'')} \quad (5.24)$$

respectively. For the distribution (5.21) this becomes

$$\frac{1}{h}\delta(q_1 - q_2 + Q) \quad \text{or} \quad \frac{1}{h}\delta(p_1 + p_2 - P). \quad (5.25)$$

The correlation between p_1 and p_2 or q_1 and q_2 respectively has entirely disappeared.

If the state of 2 is entirely ignored, the distribution of the infringed state of 1 can be found from (5.22) by integration over p_2 and q_2 with density function $1/h$. This gives

$$\frac{1}{h} e^{-\frac{i}{\hbar}q_1(P'-P'')} e^{\frac{i}{\hbar}p_1(Q'-Q'')} e^{-\frac{i}{\hbar}\frac{P'Q'-P''Q''}{2}}. \quad (5.26)$$

For the distribution (5.21) the result is $1/h$, the infringed state is entirely undetermined (the normalization can be understood from (5.26)). A measuring result $q_2 = q_{2\mu}$ or $p_2 = p_{2\rho}$ selects from (5.23) or (5.24) for 1 the distribution

$$\frac{1}{h}\,\delta\!\left(q_1 - q_{2\mu} + \frac{Q' + Q''}{2}\right) e^{-\frac{i}{\hbar}\,(q_1 + q_{2\mu})\frac{P' - P''}{2}}\, e^{\frac{i}{\hbar}\,\left(p_1 - \frac{P' + P''}{2}\right)(Q' - Q'')} \quad (5.27)$$

or

$$\frac{1}{h}\,\delta\!\left(p_1 + p_{2\rho} - \frac{P' + P''}{2}\right) e^{\frac{i}{\hbar}\,(p_1 - p_{2\rho})\frac{Q' - Q''}{2}}\, e^{-\frac{i}{\hbar}\,\left(q_1 + \frac{Q' + Q''}{2}\right)(P' - P'')}. \quad (5.28)$$

For (5.25) this gives

$$\frac{1}{h}\,\delta(q_1 - q_{2\mu} + Q) \;\; \text{or} \;\; \frac{1}{h}\,\delta(p_1 + p_{2\rho} - P). \quad (5.29)$$

Also in this example, in which all distribution functions derived from (5.21) are non-negative definite, it is already the particular part of the immediate integration over half of the parameters even in a single measurement, which does not fit into an interpretation of the 1st kind.

These few attempts and failures to carry through a genuine statistical description of the 1st kind S^1 may suffice to illustrate the intention and troubles of such a conception.

REFERENCES

1) J. v. N e u m a n n, Mathematische Grundlagen der Quantenmechanik, Berlin 1932; New York 1943.
2) H. W e y l, Z.Phys. **46**, 1, 1927; Gruppentheorie und Quantenmechanik, Leipzig 1928.
3) W. H. F u r r y, Phys. Rev. (2) **49**, 393, 476, 1936.
4) E. S c h r ö d i n g e r, Proc. Camb. Phil. Soc. **31**, 555, 1935; **32**, 446, 1936; Naturw. **23**, 807, 823, 844, 1935.
5) A. E. R u a r k, Phys. Rev. (2) **48**, 446, 1935.
6) L. F. v. W e i z s ä c k e r, Z. Phys. **70**, 114, 1931.
7) A. E i n s t e i n, B. P o d o l s k y and N. R o s e n, Phys. Rev. (2) **47**, 777, 1935.
8) N. B o h r, Phys. Rev. (2) **48**, 466, 1935.
9) G. T e m p l e, Nature **135**, 957, 1935; H. F r ö h l i c h and E. G u t h, G. T e m p l e, Nature **136**, 179, 1935; R. P e i e r l s, Nature **136**, 395, 1935.
10) E. W i g n e r, Phys. Rev. (2) **40**, 749, 1932.

STELLINGEN

I

De door H y l l e r a a s afgeleide betrekking voor verwisseling van twee elementair-quantummechanische operatoren f en g, waarbij f niet van de coordinatenoperatoren q_1, q_2,.... of $\dot{g}$ niet van de impulsoperatoren p_1, p_2,.... afhangt, kan worden uitgebreid tot de voor willekeurige elementair-quantummechanische operatoren geldende betrekking

$$\frac{\hbar}{i} \left(\frac{\delta}{\delta q_1} \frac{\partial}{\partial p_1} - \frac{\delta}{\delta p_1} \frac{\partial}{\partial q_1} + \frac{\delta}{\delta q_2} \frac{\partial}{\partial p_2} - \frac{\delta .}{\delta p_2} \frac{\partial}{\partial q_2} + \cdots \right)$$

$$fg = \dot{g}e \qquad\qquad\qquad f,$$

waarin differentiatie met δ naar links, met ∂ naar rechts wordt uitgevoerd.

E. A. H y l l e r a a s, Z. Phys. **74**, 216, 1932.

II

In de theorie van P a u l i en W e i s s k o p f voor deeltjes met B o s e - E i n s t e i n statistiek zijn de negatieve toestanden stilzwijgend vermeden op een manier, geheel gelijkwaardig met die, waarop D i r a c ze uitdrukkelijk vermijdt in zijn gatentheorie voor deeltjes met F e r m i - D i r a c statistiek.

W. P a u l i und V. W e i s s k o p f, Helv. Phys. Acta **7**, 709, 1934.
P. A. M. D i r a c, Proc. roy. Soc. London A **126**, 360, 1930; **180**, 1, 1942.
W. P a u l i, Rev. Mod. Phys. **13**, 203, 1941; **15**, 175, 1943.

De formele bewijzen voor het verband tussen spin en statistiek veronderstellen dat de scheppings- en vernietigingsoperatoren ψ en $\psi^\dagger$ elkanders geadjungeerde zijn. De redenering dient te worden omgekeerd, zodat de veronderstelde eigenschap wordt afgeleid uit het experimenteel vastgestelde verband tussen spin en statistiek.

W. Pauli, Phys. Rev. (2) **58**, 716, 1940.

Deeltjes met Einstein-Bose statistiek kunnen niet eenvoudig met een één-tijdtheorie worden behandeld.

III

Evenals de klassieke electrodynamica met puntladingen volgens Dirac, kan ook de quantumelectrodynamica in de veronderstelling van een volledig absorberend heelal worden geformuleerd als een unitaire ladingstheorie.

P. A. M. Dirac, Proc. roy. Soc. London A **167**, 148, 1939.
H. J. Bhaba and Harish-Chandra, Proc. roy. Soc. London A **183**, 134, 1944; **185**, 250, 1946.
J. A. Wheeler and R. P. Feynman, Rev. Mod. Phys. **17**, 157, 1945.

IV

Voordat men de vraag of in de dispersieformule van een electrische geleider de Lorentz-Lorenz correctie wel of niet op de geleidingselectronen moet worden toegepast kan trachten te beantwoorden, moet het oorspronkelijke begrip van deze correctie worden uitgebreid. Geeft men dit begrip een zodanige uitbreiding, dat de vraag zin krijgt, dan zal met de ter beschikking staande klassieke of quantummechanische theoriën een afdoende beantwoording zeer moeilijk of zelfs onmogelijk zijn.

C. G. Darwin, Proc. roy. Soc. London A **146**, 17, 1934.
R. Kronig and H. J. Groenewold, Physica **1**, 255, 1934.
A. H. Wilson, Proc. roy. Soc. London A **151**, 274, 1935.

In de dispersieformule voor $n^2 - 1/1 + \Theta (n^2 - 1) (0 \leqslant \Theta \leqslant 1)$ treedt zowel klassiek als quantummechanisch en zonder verdere restricties of correcties alléén bij electrische geleiders en dan nog alléén in de Sellmeyer vorm ($\Theta = 0$) een nulfrequentie op welke dan aan de geleidingselectronen of ionen kan worden toegeschreven.

C. G. Darwin, l.c.

V

De door P e s h k o v in vloeibaar helium II gemeten warmtegolven kunnen met de theorie van K i r c h o f f in het geheel niet worden verklaard.

G. S t a n l e y S m i t h, Nature **157**, 200, 1946.
E. G. R i c h a r d s o n, Nature **158**, 296, 1946.
F. H. v. d. D u n g e n, Bull. Acad. Belgique (Classe de Sciences) (5) **19**, 1180, 1930.
H. J. G r o e n e w o l d, Physica **6**, 303, 1939.

Een ook maar qualitatief bevredigende verklaring van transportverschijnselen in helium II is nog niet gegeven.

De vraag of de λ-overgang in vloeibaar helium een verschijnsel van E i n s t e i n condensatie is kan in beginsel zowel theoretisch als experimenteel worden opgelost. De moeilijkheden van een theoretische beslissing zijn van interne aard. De experimentele beslissing daarentegen is thans technisch uitvoerbaar, maar stuit op hoge kosten en geheimhouding van wetenschappelijke gegevens.

VI

De bij het cascadeproces van neutronen in paraffine gebruikelijke waarschijnlijkheidsbeschouwingen zijn onbevredigend en voor verbetering vatbaar.

E. A m a l d i e.a., Proc. roy. Soc. London A **149**, 522, 1935.
G. C. W i c k, Phys. Rev. (2) **49**, 192, 1936.
E. U. C o n d o n and G. B r e i t, Phys. Rev. (2) **49**, 229, 1935.
S. G o u d s m i t, Phys. Rev. (2) **49**, 406, 1935.
H. A. B e t h e, Rev. Mod. Phys. **9**, 69, 1937.

In het niet-thermische gebied van de door R a i n w a t c r en H a v e n s gegeven tabel moet het aantal botsingen met 1 vermeerderd worden, terwijl de benodigde tijden een factor variërende tussen 1 en 2,5 maal te groot zijn opgegeven. Het vermoeden van B a k e r en B a c h e r, dat na ongeveer 10 μ sec het evenwicht voor de niet-thermische neutronen is ingesteld, kan worden gerechtvaardigd.

C. P. B a k e r and R. F. B a c h e r, Phys. Rev. (2) **59**, 332, 1941.
R. F. B a c h e r, C. P. B a k e r and B. D. M c D a n i e l, Phys. Rev. (2) **69**, 443, 1946.
J. R a i n w a t e r and W. H a v e n s, Phys. Rev. (2) **70**, 136, 154, 1946.

VII

Het door B e t h aan T a r s k i ontleende voorbeeld van een axiomastelsel, dat niet-strijdig is en geen interpretatie toelaat, geeft aanleiding tot bedenkingen, de gevolgde bewijsvoering tot tegenspraak.

E. W. B e t h, Inleiding tot de wijsbegeerte der wiskunde, Nijmegen–Utrecht, Antwerpen–Brussel 1940 (blz. 114 e.v.).

VIII

Tussen de opvattingen van B o h r en die van S c h r ö d i n-g e r over de betekenis van de quantummechanische onbepaald-heid bestaat ook inzake de rol, die deze in de biologie speelt, een controversie. In het laatste geval zijn de opvattingen van S c h r ö-d i n g e r minder speculatief dan die van B o h r.

N. B o h r, Naturwiss. **21**, 245, 1933.
E. S c h r ö d i n g e r, What is Life?, Cambridge 1945.

IX

De ontwikkeling van de ervaringswetenschappen zou ernstig ge-schaad worden, wanneer metaphysische elementen (in de logisch-empiristische zin) bij voorbaat uit het scheppend proces van het stellen en oplossen van problemen zouden worden uitgebannen.

Het optreden van metaphysische elementen in problemen der ervaringswetenschappen wijst er op, dat de probleemstelling niet ten volle doorzien of het probleem niet geheel opgelost is.

X

De late sociale bewustwording van de wetenschappelijke onder-zoekers hangt samen met hun plaats in de maatschappelijke struc-tuur. Dat deze plaats nog weinig is opgehelderd hangt samen met hun late sociale bewustwording.